Cracking the Code of Our Physical Universe

Matthew M. Radmanesh, Ph.D.

Why This Book?

What sets this book apart is the fact that it is not just another science book describing scientific facts and phenomena! It would surely be redundant since that task has been done many times over with much more elegant prose and brighter narrators.

In this book, for the first time we have undertaken the task of breaking the code of any piece of matter or natural phenomena; whether it is an atom, a quantum occurance, a planet, a galaxy, or any other perceivable thing. It covers any natural phenomena ever discovered or one that will be unravelled by the future pioneers in their respective fields.

This book provides the trail map of any and all things that man has discovered and shows how their codes were cracked. The list of discoveries is endless but prominent amongst them are the discovery of fire, elecricity, magnetism, laws of motion, the solar system and planets, so on and so forth.

This book goes beyond just pure science since it fuses philosophy with science. It actually makes science a subset of philosophy, or more precisely, applied philosophy. Just like the light phenomenon, which was made to be a subset of the field of electricity by James Clerk Maxwell, revolutionizing our technical world, so does this book by bringing a new era of incredible developments for mankind!

For more information please visit:

www.KRCbooks.com

More Praise for "Cracking the Code of Our Physical Universe"

"Not since Stephen Hawking has anyone so captured the essence of the Universe and all of its simplicity and complexity! A delightful must read for anyone interested in the foundations of our Universe!"

—Dr. Charles K. Alexander,
Dean of Engineering, Cleveland State University

"A magnificent work, which is the most inspiring book in one complete and easy to understand package"

—Dr. Bob Salajekeh,
Professor of Mechanical Engineering, Kettering University

"A unique and fascinating approach to the universe, this book is a must for anyone who's interested in exploring the mysteries of our complex world."

—Dr. Brenda Timmerman,
Computer Scientist, California State University, Northridge

"The complex ideas of our universe are presented brilliantly for a complete and practical understanding."

—Theresa Locke,
Chief Editor, The LATEST Magazine

"A thought provoking treatment of the complex issues our universe presents to us in an easy to understand manner."

—Dr. Jeff Wiegley,
Computer Scientist, California State University, Northridge

" 'Knowledge Economy' is the key element for our future success in the universe. This book introduces the tools to master this area ."

—Dr. Mo Torfeh,
Professor of Electrical Engineering, Kettering University

"This book is loaded with novel ideas and insights; an easy-to-understand presentation of the author's unique vision of the universe."

—Dr. John Noga,
Professor of computer science, California State University, Northridge

"Finally, a powerful book that translates science into a common language, which can increase your understanding of the world."

—Robb Gray,
President, First Pacific Financial, Inc.

Cracking The CODE of Our Physical Universe

The Key to a Whole New World of Enlightenment and Enrichment

Matthew M. Radmanesh, Ph.D.
Professor of Electrical
& Computer Engineering,
California State University, Northridge

authorHOUSE
1663 LIBERTY DRIVE, SUITE 200
BLOOMINGTON, INDIANA 47403
(800) 839-8640
www.authorhouse.com

This book is a work of non-fiction. Unless otherwise noted, the author and the publisher make no explicit guarantees as to the accuracy of the information contained in this book and in some cases, names of people and places have been changed to protect their privacy.

First published by AuthorHouse 05/18/06

ISBN: 1-4259-1600-7 (e)
ISBN: 1-4259-1599-x (sc)
ISBN: 1-4259-1598-1 (dj)

Library of Congress Control Number: 2006904633

Printed in the United States of America
Bloomington, Indiana

This book is printed on acid-free paper.

Front Cover Image*: The design on the front cover shows the graceful, winding arms of the majestic spiral galaxy M51 (NGC 5194) appearing like a grand spiral staircase sweeping through space. They are actually long lanes of stars and gas laced with dust. This sharpest-ever image of the* **Whirlpool Galaxy** *(located 31 million light-years away), taken in January 2005 with the Advanced Camera for Surveys aboard NASA's Hubble Space Telescope, illustrates a spiral galaxy's grand design, from its curving spiral arms, where young stars reside, to its yellowish central core, a home of older stars. The image is 14,000 light-years across.*

Back Cover Image*: The design on the back cover shows* **Star-Birth Clouds** *in M16. This eerie, dark structure, resembling an imaginary sea serpent's head, is a column of cool molecular hydrogen gas (two atoms of hydrogen in each molecule) and dust that is an incubator for new stars. The stars are embedded inside finger-like protrusions extending from the top of the nebula. Each "fingertip" is somewhat larger than our own solar system. The pillar protrudes from the interior wall of a dark molecular cloud. It is part of the "Eagle Nebula" (also called M16), a nearby star-forming region 6,500 light-years away in the constellation Serpens.*

Dedicated to the

Best of Mankind,

Whose Spiritual Wisdom

Has Lit Up

This Forlorn Corner

of Our Galaxy!

Contents

Foreword

Not since Stephen Hawking has anyone written with such brilliance, clarity of thought, and in a way any of us can understand!

Science and engineering have always needed outstanding communicators who communicate without all the excess scientific terminology and complex grammar that even a professor of English would have a hard time understanding.

Dr. Radmanesh not only captures the technical complexity of scientific thought, he marries it to complex philosophical thought and serves them up like a delicious club sandwich. He extends the boundaries created by Hawking and finally brings the theoretical physics into the real engineering world to complete the story.

I highly recommend this book to anyone interested in our world and our future! I cannot wait to purchase a copy for my wife and one for my mother.

Dr. Charles K. Alexander, Fellow IEEE, P.E.
Dean, Fenn College of Engineering,
Cleveland State University;
CEO of the IEEE (1997)

Preface

Recently the author published a popular book entitled "The Gateway to Understanding: Electrons to Waves and Beyond" which was well received by the scientific community. At the time it appeared that some of the main goals set forth by the author in writing the book were fulfilled. However, there were areas that were not clearly addressed in that work and needed further elucidation.

The need for such a further elucidation is what led to the creation of this second book and is where it takes off. It is the sequel or the second installment to the first book and intends to undercut areas that were not illuminated in the first volume.

The present work is the culmination of many years of study, observation, and pondering on the dilemmas and enigmas of the physical universe and their origin and the resultant understandings that were extracted from this sophisticated and at times incomprehensible arena.

It is an interestingly uncommon book written primarily for the technical as well as the non-technical man. It is intended to serve several classes of our society

a) The technically versed individuals,
b) The interested but non-technical individuals,
c) The professional scientists.

This book will also surely serve also an important class of our society—the technical inventors who are looking for inspirations and new ideas to imbue them with enough understanding to finalize and materialize their thoughts into reality.

It is also written for the average man who may or may not be technically versed and yet desires to learn about the universe at large, or the technical world in his immediate surroundings. It is intended to lift the aura of "black magic" surrounding the world of sciences, to enlighten and demystify the subject of sciences in the minds of ordinary individuals.

The Importance of Work

Everyone, in today's society, is struggling with this dominant and imposing thing called the physical universe and strives toward a higher understanding of its inner workings. Yet most books present the basic concepts with so much complexity and filled with so many mathematical equations that the general public has given up on the subject, choosing to retire to the sideline as spectators. In other words, their hopes have been dashed aside and their dreams of a higher understanding have not been fulfilled in any of these books.

Within the confines of this book, one is given a chance for the first time to take an in-depth look and inspect first-hand the code of one of the most enigmatic universe that has ever been constructed. Its dominance and imposing characteristics in all aspects of our existence is truly remarkable. The basics are stated in simple terms and clear explanations express the powerful principles lucidly and dynamically, providing an unforgettable impression in the reader's mind.

This is a new approach unmatched in any extant text today. The discovery of these fundamentals has had a huge impact on our current world and has truly made our scientific arena a bright beacon of hope with a renewed interest in understanding our physical universe. This work has created, in very simple terms, a "unified theory" about the two distinct concepts: physical and thought universes.

Finally, this work lays out the milestones, so that the scientist as well as the non-technical individual can travel on this adventurous road and be able to formulate and develop the code that starts to crack open the "Material or Physical Universe" and ends up with the key to the kingdom of the "Thought Universe."

The Author's Goals

The author intends to bring forth a milestone achievement that can be summed up as:

To bring about a public awareness of how the universe affects us in many ways on a constant basis and how a basic knowledge of the subject not only can lead to the cracking its code but also gaining a higher level of causation over it.

The non-technical reader will be invited to examine a series of basic materials that will enable him/her to understand his immediate universe far better than ever before. He/she will be exposed to materials of considerable significance, which surely would open up the gates of knowledge along with a wider horizon of understanding.

Any communications in the way of a healthy criticism and/or correction are welcome. Moreover, the author considers it one of the most rewarding things to have others grasp the materials in all of their simplicity and increase their own potential survival in this universe and help others to achieve their goals. In the process, this helps make Man take control of his own destiny, without being shackled by the chains of higher authority or superstition.

Therefore, in order to improve the quality of this work, the author would like to have all comments or suggestions be sent directly to:

Dr. Matthew M. Radmanesh
18111 Nordhoff Street,
Department of Electrical and Computer Engineering,
California State University, Northridge, California 91330.

Or email to: **matt@csun.edu**

You can also check out these related websites for more information:
www.KRCbooks.com
www.csun.edu/~matt

Matthew M. Radmanesh, Ph.D.

Acknowledgements

First and foremost, special thanks are due to Dr. L. Ron Hubbard, a true genius, whose pioneering works in several vital arenas, brought many new discoveries to the forefront and inspired the author to write the original manuscript to formulate the methodology of how the code of the physical universe could be cracked.

The author's special gratitude also goes to a dear friend, Robb Gray, who helped him in many ways to keep his thoughts focused and to provide much needed support during this intense project. Thanks are also due to a highly-valued and special individual, Jaime Rodriguez, who simplified many challenging aspects of this work.

The author would like to further thank many of his professional colleagues, particularly Dr. Asad Madni (CEO/COO of BEI Technologies, Inc.), one of his best friends/colleagues, Dr. Charles Alexander (Dean, Cleveland State University), Dr. S. T. Mau (CECS Dean), Dr. Nagi El Naga (ECE Dept. Chair), Dr. Tom Mincer, Dr. Brenda Timmerman, Dr. Jeff Wiegley, Dr. John Noga, Dr. Tim Fox, Dr. Robert Burger (California State University, Northridge, CA) dear friends at CSUN, Philip Arnold (Agilent), a good friend, Dr. George Haddad & Dr. C. M. Chu (University of Michigan, Ann Arbor, MI), the early mentors in Michigan, Dr. M. Torfeh and Dr. B. Salajekeh (Kettering University, Flint, MI). Their support and collegiality through the years is definitely appreciated.

Finally, the author's deep gratitude belongs to his lovely wife, Jane Marie, and his lovely son, Jeremy William, for making life soothing and sweet during this power-packed project and to his parents, Mary and the late Dr. G. H. Radmanesh, for their true love and unconditional support.

Matthew M. Radmanesh, Ph.D.
Dept. of Electrical and Computer Engineering,
California State University, Northridge,
May 2006

What Sets This Book Apart

What sets this book apart is the fact that it is not just another science book describing scientific facts and phenomena. It would surely be redundant since that task has been done many times over with much more elegant prose and brighter narrators.

Here is a book where, for the first time, we have undertaken the task of breaking the code of any piece of matter or natural phenomena; whether it is an atom, a quantum occurrence, a planet, a galaxy, or any other perceivable thing. It covers any natural phenomena ever discovered or one that will be unraveled by their future pioneers in their respective fields.

"*To be philosophy's slave is to be free.*"

—Seneca Wallace (5 - 65 AD)
Roman Philosopher

This book lays the trail map of any and all things that man has discovered and shows how to bring about new ones. The list of discoveries is endless. Prominent amongst them are the discovery of fire, electricity, magnetism, laws of motion, the planets, the solar system so on and so forth.

Just like the light phenomenon, which was made to be a subset of electricity by James Clerk Maxwell, thus revolutionizing our world, so would this book by bringing on a new era of incredible developments for mankind!

This book is the road map, which leads to the final disentanglement of the universes and makes one stand at cause over any and all of the created universes to change the course of their creation!

"*Superstition sets the whole world in flames; philosophy quenches them.*"

—Voltaire (1694-1778)
French Philosopher

Why Should I Read This Book?

Cracking the code of the "*Physical Universe*" and deciphering its many mysteries is a daunting task that has doggedly been pursued for many generations, since the beginning of recorded time on this planet. The design and origin of the code goes back to trillions of years ago, long before the appearance of Homo sapiens (Modern Man), as a species on earth, which happened about one hundred thousand years ago.

The main features of the "code" and its many complexities could not have possibly been cracked before the late 1800s, when the laws of electricity, magnetism and electronic waves were finally formulated and put forth as a unified theory by several prominent scientists.

There have been much speculations, wild theories and complex thoughts about the nature and the structure of the code, and many philosophical approaches have been developed through the ages in an effort to resolve a piece of the puzzle in the hope of understanding the rest of the puzzle through a process called "inductive logic."

"Tell me and I'll forget; show me and I may remember; involve me and I'll understand."

—A Chinese Proverb

Thus, the search for understanding a puzzle of enormous proportion called "The Physical Universe," has been in the works for centuries, if not millennia.

If we live on this planet and go about our daily world activities in complete disregard of this imposing universe, we will certainly have lots of trouble enhancing our own survival potential as well as our progenies and symbionts (i.e., life forms that are mutually dependent, e.g., pets, plants, etc.).

However, if we engage upon a campaign in understanding the intricacies and the inner workings of the physical universe, then maybe we can formulate our own version of the code and finally achieve a total mastery of one of the most enigmatic universes ever created.

"Every master was once a disaster."

—T. Harv Eker (circa 1960-)
Canadian Educator, Writer, and Motivational Speaker

So, we need to know the "code" to the physical universe in all of its ramifications to understand how it works if we intend to make a wholehearted effort in resolving and disentangling our own universe from this utterly amazing and extremely dominant one.

This campaign starts here and now with this book, so read on my fellow traveler since you are going to be on a journey— a "road to discovery and knowledge," if you will, and will come away at the end with a heightened awareness of many facets of our universe, hopefully leading to a more joyful, causative, and rewarding life.

"***Life-transforming ideas have always come to me through books.***"

—Bell Hooks (1952-)
Distinguished Professor of English

Cracking the Code.....

Since time immemorial, Man has been on a long and relentless search for the answers to the riddles and puzzles of this universe. It has been a long search, and many a great scientist and philosopher have been at work to crack the code by revealing an ultimate solution in a simple and understandable fashion.

There have been bits and pieces to this puzzle discovered through the millennia, particularly, over the last one hundred years.

What we are attempting to undertake in this work is what has embroiled many through the past ages and still beckons as unreachable to some today.

Cracking the code depends on several key factors, especially the code disintegrator, which plays a major role in it as will be seen shortly. Cracking the code and its discovery will undoubtedly bring about a clarification of the material side of life and point out its origin as it has evolved to be what is today—a complicated and uncertain universe.....

PART I

A Glimpse of Our Immense Universe

Our Inner Resources

Before we embark upon this journey, let us take stock of our own mental resources, the non-physical aspects of our existence with respect to the physical universe, with which we are in constant association. We need to grasp our own position in space, where we are at the moment, our wherewithal, our spiritual and mental resources, and above all, what we are doing here as a species, before we can understand the essence of our universe.

Upon close observation we can uncover some amazing discoveries about ourselves. We quickly learn that we are fortunate indeed! We find ourselves in a body form, a low combustion carbon–oxygen engine, a physically feeble entity, which cannot tolerate much force. It seems that we have been left alone in a very desolate place.

We are an inhabitant of a thin crust of a minor planet of an inconsequential star in just one of the millions of galaxies in the universe. However, ***we are born with the ability to observe, to communicate, to reason, to understand our surroundings, to thus be able to solve our essential survival problems, and moreover, to find out our relationship to the physical universe with which we are intimately associated.***

"Ability will never catch up with the demand for it."

—Malcolm Forbes (1919 - 1990)
American Author & Publisher

We do not merely exist or survive by tooth and claw or by rote and instinct basis like the rest of the earth's creatures. We ask the why and how of things. To find answers to the mysteries and riddles of this physical universe, we look into and under and behind them. We have the ability to ponder problems of existence and survival and come up with workable solutions, if not perfect!

"There are no limitations to the mind except those we acknowledge."

—Napoleon Hill (1883-1970)
American Author

We are not amoebas inhabiting earth, which are floating aimlessly in the seemingly boundless space of the physical universe. We are not matter but the life force that imbues life into matter and energizes the physical body. We, being the product of the two, are complex beings that have an understanding and an awareness of the physical universe, a high propensity as well as an ever-increasing capacity to shape it to our own needs and desires.

"When every physical and mental resource is focused, one's power to solve a problem multiplies tremendously."

—Norman V. Peale (1898-1993)
American Writer

While being in a body, we cannot begin to grasp the magnitude of the created space physically surrounding us; however, intellectually we have grasped this concept and its origin. Our constant search for the answers and use of very powerful instruments, have enabled us to probe deeper and deeper into the cosmos. Our computation at this time indicates that we live in a physical universe with an approximate diameter of 8,000,000,000 light years ($8x10^9$ or 8 billion light-years, ly), equivalent to a radius of about 4 billion light-years.

Now, considering a light year to be the distance that light, at the speed of 186,000 miles per second, travels in one year we find that this is an unfathomably large number relative to the size of a human body. Considering the existing modes of transportation, their speed, and human life span, we find these distances to be impossible to travel!

"With realization of one's own potential and self-confidence in one's ability, one can build a better world."

—Dalai Lama (1935-)
Head of the Tibetan Buddhists, 1989 Nobel Peace Prize Winner

Our Galaxy

In a sea of enormous space, there are billions of star atolls, (i.e. the galaxies) each containing billions of stars, clouds of hydrogen and dust. While the average size of a star is about the size of our sun, which fits a million earths within it; however, there are stars that could fit a million suns.

"And where in vastness of all this is our galaxy?" one may ask. In an aggregate of twenty galaxies known as "The Local Group," there is one called "The Milky Way," an elliptical pinwheel of 100 billion stars. At a position of about 33,000 light years from the Milky Way's center (about 17,000 light years from its edge), among the countless stars in just one of the spiral arms is our sun as shown on the next page.

The time frame and the distances involved between stars and planets are so large compared to our average life span, or the speeds and the distances encountered on earth, that it makes it difficult to grasp the totality of this incredible scene and thus develop a reality toward it or be able to put all of these into a proper frame of reference.

" Human beings, vegetables, or cosmic dust, we all dance to a mysterious tune, intoned in the distance by an invisible player."

—Albert Einstein (1879-1955)

German Born American Physicist, Nobel Prize for Physics in 1921

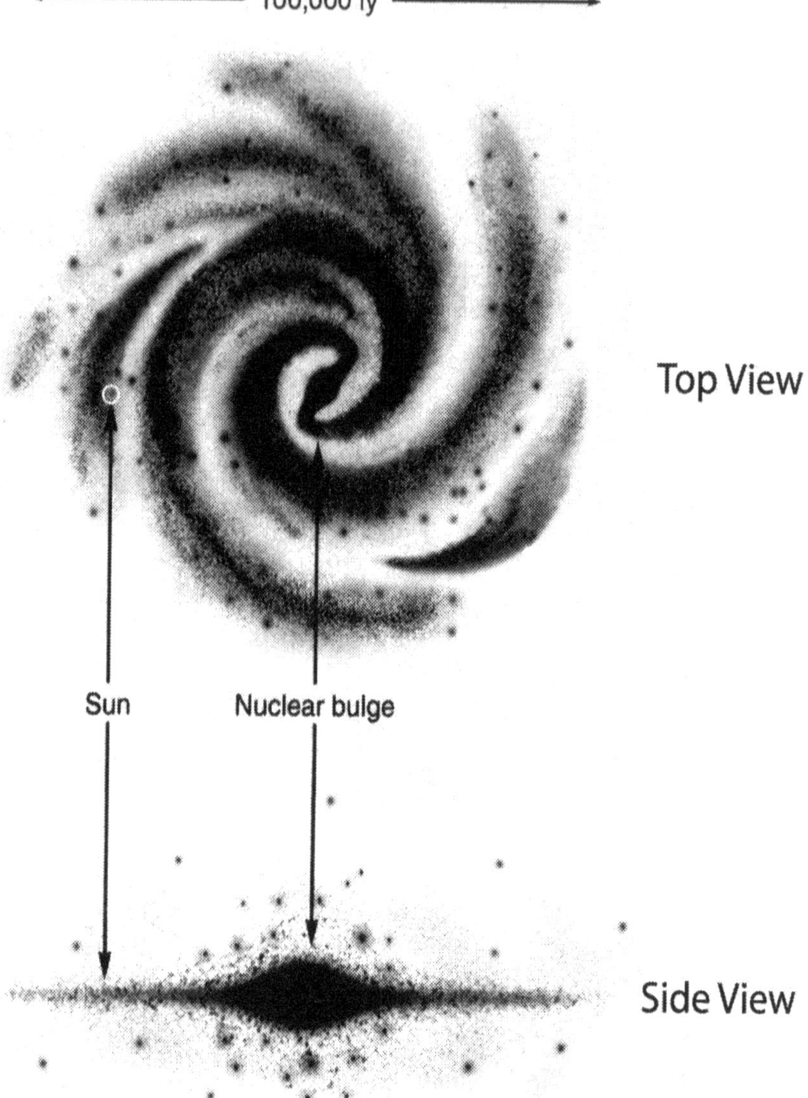

The Milky Way Galaxy Spanning 100,000 Light-Years (Lys) And The Position Of Our Sun In It.

Our Solar System

If we imagine a gigantic elliptical pinwheel spanning approximately 100,000 light years in diameter, containing billions of stars, planets and moons on each of the spiraling arm, we are beginning to get a good idea about the size of our galaxy. Overall our galaxy is home to 100 billion stars. Our solar system is considered to be on the outer edge, relatively far from the center of the galaxy (approximately 2/3 away or 33,000 light years from the center) on one of the spiraling arms. From earth, we cannot observe the Milky Way's glowing center, since it is obscured by clouds of interstellar dust, black holes and so on. For this reason our "window to the center of our galaxy" is blocked by drifting cosmic matter. However, the "window to the whole physical universe" is at right angles to the plane of our galaxy, which is an almost a two-dimensional plane corresponding to the plane of the pinwheel face. This is because the view is quite open and least unobscured by the interstellar dusts and cosmic matter and billions of stars.

Among the billions of stars in our galaxy, there are believed to be millions of solar systems, much similar to our own. But to date, Man has not been able to observe another such group of cold heavenly bodies, which are lit by the reflected light of a central sun. It appears to be a close-knit family of nine planets but it is not. It actually extends 3.5 billion miles in space.

To get an idea of relative size of the solar system, let us build a *conceptual model* by representing the sun as a soccer ball. Then on this basis, earth would be a small marble 200 feet away (the distance from earth to sun is commonly referred to as one *Astronomical Unit* (AU), where 1 AU$\approx$1.5x10^8 km), whereas the furthest planet, Pluto, would be another small marble 7,900 ft beyond. Since there are 5,280 feet per mile then Pluto is 1.5 miles away relative to earth, making approximately a relative ratio of 40 to 1 in distance from the sun. In other words there are vast expanses of space existing between planets as shown below.

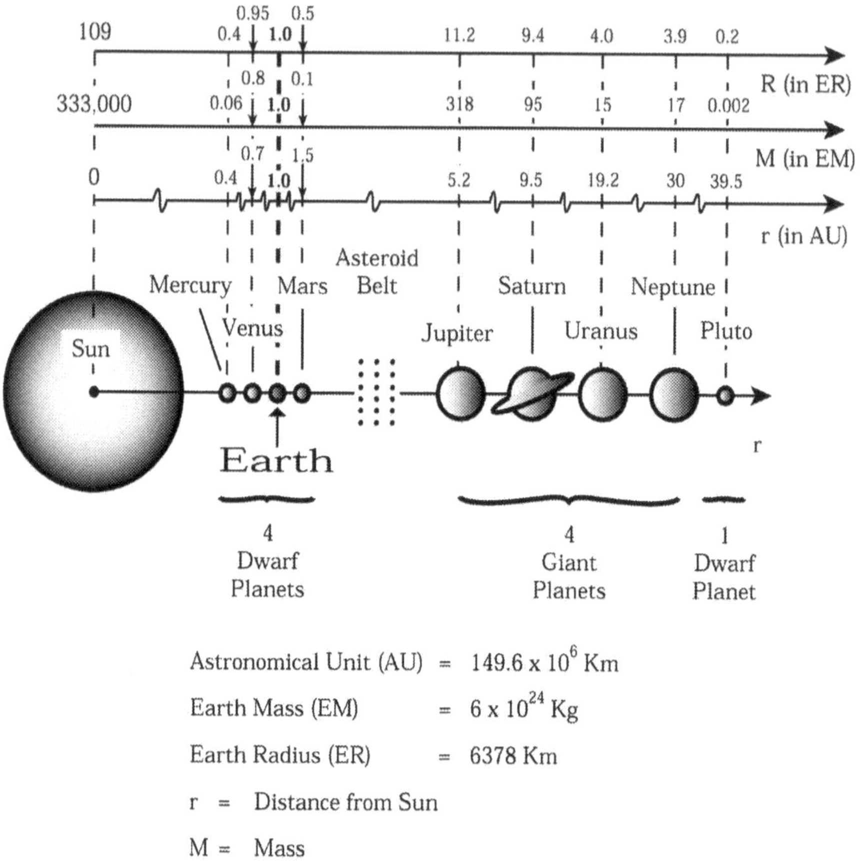

Astronomical Unit (AU) = 149.6×10^6 Km

Earth Mass (EM) = 6×10^{24} Kg

Earth Radius (ER) = 6378 Km

r = Distance from Sun

M = Mass

R = Radius

Note: Diagram is not drawn to scale.

The Solar System.

Starting from sun, there are four inner dwarf planets: ***Mercury*** (0.39 AU), ***Venus*** (0.72 AU), ***Earth*** (1.0 AU) and ***Mars*** (1.52 AU). Beyond this group of tiny planets, lies a belt of *asteroids*. Asteroids are very small planets with a maximum diameter of 490 miles and for this reason they are also called planetoids.

Beyond these are four giant planets: ***Jupiter*** (5.20 AU), ***Saturn*** (9.54 AU), ***Uranus*** (19.20 AU), and ***Neptune*** (30.06 AU). Finally, beyond these planets lies another dwarf planet, ***Pluto*** (39.44 AU), which is believed to have been once a moon of Neptune. There are also a total of 31 moons and satellites orbiting these planets.

There are ***Flying Comets***, which appear to be fiery interplanetary masses scattering their embers as they move. In actuality these are heavenly bodies having a star-like nucleus with a luminous mass around it and usually a long luminous tail that points away from the sun. These bodies usually follow an elliptical or parabolic orbit around the sun.

There are also ***Meteoroids***, which are many small solid bodies traveling through the outer space, which become visible and can be seen as meteors. A ***Meteor***, loosely called a ***Shooting (or Falling) Star***, is the flash and streak of light and the ionized trail occurring when a meteoroid is heated by its entry into the earth's atmosphere. The part that survives passage through the atmosphere and falls to the earth's surface, as a mass of metal and stone, is called a ***Meteorite***.

"I don't think the human race will survive the next thousand years, unless we spread into space. There are too many accidents that can befall life on a single planet. But I'm an optimist. We will reach out to the stars."

—Stephen Hawking (1942-)
English Physicist

Our Sun

The Intense light of the sun is the fire produced by the nuclear conversion of Hydrogen into Helium. The surface temperature is about 11,000 °F, but the interior temperature is millions of degrees higher. Periodically, storms erupt from the surface and projections of fire are hurled millions of miles into space, bombarding the earth's atmosphere with streamers of high-energy radiation.

"The glorious lamp of heaven, the radiant sun, is nature's eye."

—John Dryden (1631-1700)
British Poet, Dramatist

Our sun provides an eternal source of energy for all life forms, from vegetable kingdom to animal kingdom, which includes mankind. Its constant supply of energy in the form of radiation of sunlight makes life possible and tenable on earth.

Our sun is the central stable datum upon which the entire solar system has come together in a neat package of planets and moons. Due to its enormous size and weight relative to all other nearby heavenly bodies, our sun has created perforce an enormous level of stability in this corner of galaxy.

"Hope is like the sun, which, as we journey toward it, casts the shadow of our burden behind us."

—Samuel Smiles (1812-1904)
Scottish Author

At first glance, the solar system appears to be a friendly family of planets. It consists of one yellow sun, nine planets, 31 moons and satellites and thousands of asteroids, comets and meteors. It is a series of planets packed neatly together, but we find it to be quite the contrary.

The face of Mercury nearest to the sun is an eternally scorched desolation due to its proximity to sun. Venus is veiled in a mystery of clouds. Mars is intriguing and the most promising for future explorations. The cold and barren giants (Jupiter, Saturn, Uranus, and Neptune) are surrounded by deadly gases. Thus, upon further and close examination of its occupants, we find it a ghastly and forbidding family!

Grasping the panoramic view of the physical universe as discussed above, is essential in gaining a solid foundation for understanding the enormity of our universe, which we may be oblivious due to our own physical handicap of being bound to a body form, which is held down to the earth's surface due to high gravitational fields.

Since there are many other galaxies, which contain a large number of planets, stars, suns, and nebula and so on, therefore when we say "The Universe," we are dealing with an enormous mass of bodies and vast spaces between them. Nevertheless, this is the universe through a small portion of which we exist and communicate to others on a physical plane.

We have to progressively narrow our attention to a smaller and smaller area of the physical universe to find where we live physically and understand the vantage point from which we view the universe at large, or sciences in particular. We will parallel a similar pattern in this work, so the presented materials strike the correct chord of reality in the reader's mind for easy grasp!

"To him whose elastic and vigorous thought keeps pace with the sun, the day is a perpetual morning."

— Henry David Thoreau (1817-1862)
American Essayist, Poet and Philosopher

Earth

In this vastness of space, high speed energy particles and enormous numbers of masses of all shapes and sizes, which have persisted over many trillions of years, we find earth —a small planet on a remote solar system in the Milky Way galaxy.

Planet earth is spinning about its axis at a speed of about 1,000 miles per hour and rotating around the sun at 66,000 miles per hour. In turn, the solar system along with all other stars and planets is circling the far distant center of the Milky Way once every 200,000,000 years (200 million years) as our great galaxy, as a whole, hurdles through the space of the physical universe.

This is where our home is as we wheel through the physical universe. Our short time span of existence as a species (between 35,000 and 100,000 years) and our current state of survival on earth, is an interesting situation, which will be the subject of much ponder for many years to come.

"Now there is one outstandingly important fact regarding Spaceship Earth, and that is that no instruction book came with it."

—Richard Buckminster Fuller (1895-1983)
American Engineer and Architect

Our Home, Earth

⊶ ✶ ❁ ✶ ⊶

The Universe

It is generally understood that when we say "the universe," we actually mean "The Physical Universe." The physical universe, which happens to be the primary target of all sciences, particularly the physical sciences, is an interesting universe. It is based upon four powerful primary postulates, which in principle can be reduced to three simple postulates, more of which later.

We need to ask a basic question: "Where do we begin our study of this panoramic physical universe of such a vast expanse?" We find ourselves unable at the moment to examine the rest of the solar system let alone the Milky way Galaxy. Thus we have to draw a boundary and limit our studies to the physical universe material presented to us and available on earth at the present moment.

It should be noted that our knowledge of the center of this galaxy or other galaxies, at the moment, is based upon a second-hand level of observation, since it is purely obtained through examination of the wave emanations at many frequencies (e.g. radio waves, light waves, etc.) from the stars, and not through first-hand, close observation and direct contact.

Thus our task is to analyze the physical universe, which we are intimately in contact with through a small part of it (i.e. earth) and then try to generalize the knowledge so obtained to the rest of the solar system, to Milky Way or to all other galaxies that we know nothing of at this moment.

It is a vast and an imposing universe and controls many aspects of our existence on many facets. By observation, we note that it has a certain hidden nature to it, in that it does not reveal its inherent tendencies or its truth easily. This hidden nature asserts itself simply by being constantly out of communication with the beings who reside in it and by its propensity toward fast-paced motion on a microscopic level (if not macroscopically), which ultimately creates a complete obfuscation of truth.

The physical universe is constantly in a "***State of Change.***" The "change" takes place on an automatic level and involves changes in the shape, form, position, density and other pertinent characteristics, which are based upon the constant interaction of matter and energy on a one-directional, continuous and monotone fashion, creating a time stream called the "Mechanical Time." Physicists refer to this continuous and monotone time stream as "***The Arrow of Time.***"

"The universe contains vastly more order than Earth-life could ever demand. All those distant galaxies, irrelevant for our existence, seem as equally well ordered as our own."

—Paul Davies (1946-)
British Internationally Acclaimed Physicist and Writer

However, it has an antidote and surrenders rather easily to intelligent use of logic in a methodical fashion leading to a practical method for finding answers to any enigma including that of the physical universe. This practical method, which has been employed extensively in all of the extant physical sciences, is called the "***Scientific Methodology,***" which will be discussed later.

The physical universe is constantly asking its inhabitants to search for and try to discover its many mysteries. It has an uncontrollable propensity for motion whether macroscopically or microscopically. It could be called the "***Universe of Motion***," which has been left on auto-pilot of "change of particle's location," another name for "time."

However, it is controllable on a rather small scale, where the control factor is furnished by an exterior element called a "viewpoint." When directed properly, the auto-control factor of the physical universe can be overridden and the control on a small scale can be brought about by applying a combination of force and intelligence. A simple example of such an overriding factor is sending a wave of sufficient power and a frequency high enough to defeat the motion of the electrons in the ionospheric plasma (which acts as a mirror for any low frequency wave) and thus penetrate it and reach, for example, an interplanetary space probe, for its guidance and control.

Therefore, our study in this work has a clear direction and an exact intention; the universe that we are studying is an important universe since, a) It is the ***current playing field*** for all of our activities, and b) Its fundamental postulates and principles of operation must be well grasped if we desire a to crack its code and obtain a workable knowledge about it as a whole or any of its constituent components.

"There was no 'before' the beginning of our universe, because once upon a time there was no time."

—John D. Barrow (1952-)
Professor of Physics, Cambridge University

Universes

There is nothing wrong with an inclination on learning about the physical universe, but there is everything wrong with a total fixation or obsession with the physical universe to a point of total exclusion and unawareness of other universes and their contents.

Herein we intend to widen this narrow view of "the universe," which everyone equates mentally with the "physical universe" and present a new look on the subject. This new look will prevent this mental identification, and will give the subject of "universes" a fresh outlook, thereby establishing a whole new framework of thought and understanding of what is meant by a universe.

When we speak of the physical universe we mean a "total effect" type of universe, which is an "unthinking" universe where all observed phenomena is carried out at the level of "action-reaction" (based on Newton's third law of interaction) or on a totally automatic basis.

Examples of this type of automatic action-reaction include:

a) Revolution of earth (or other planets on their orbits) around the sun, which creates time on an automatic level.
b) Precession of a spinning earth around its axis, which creates the four seasons automatically.
c) Chemical reactions of atoms with each other, which create new substances spontaneously. For example, oxygen reacting with hydrogen creates water.
d) The gravitational field on earth, which causes an object to fall back to earth or be held on it and not escape into free space. This list could go on.

There is no thought used on any of the above or a million other "action-reaction" activities taking place on earth on a daily basis.

They are all done on an automatic and preconceived basis and without any volition or decisional power on the part of any being.

As it turns out, the concept of the physical universe being the only universe is an apparency and not an actuality. Truth be told, there are many universes and the physical universe happens to be just the most visible and the most dominant universe at the moment.

"The universe is full of magical things patiently waiting for our wits to grow sharper."

—Eden Phillpotts (1862-1960)
English Science-Fiction Writer

Universe, Definition of

Let us now broaden our understanding of the term "Universe" and define it exactly at this juncture. The word "Universe" is derived from Latin meaning "turned into one," or "a whole," and can now be defined as ***"The totality or the set of all things that exist in an area under consideration, at any one time."***

In simple terms, "Universe" denotes *an area consisting of things (such as ideas, masses, symbols, etc.) that can be classified under one heading and be regarded as one whole thing.*

Therefore, there are many other universes that one can become aware of and recognize their existence, such as the universe of waves, the universe of mathematics, the universe of symbols, the thought universe, a being's own universe, so on and ad infinitum as shown below.

Of course, sometimes a universe can be a subset of another universe or have an overlap with another universe, but that does not make it any less of a universe. As long as the above definition holds valid, it is considered to be a universe regardless of its size or other considerations.

Therefore, we need to realize that we have broadened and generalized the concept of a "Universe" and when we say "Universe" we are not necessarily referring to the "Physical Universe."

Through careful analysis and dissection of the physical universe into its various components and constituents, we can successfully isolate the exact "pattern" of its construction. Utilizing this carefully discovered pattern of construction we can now embark upon a whole new path.

This path will take us from a concept to the creation of any desired universe provided one follows a certain and exact sequence of actions. This is only possible if one knows and applies the maxim, ***"The discovered postulates and principles in the physical universe provide the essenjtial patterns and blueprints for the creation of any new universe."***

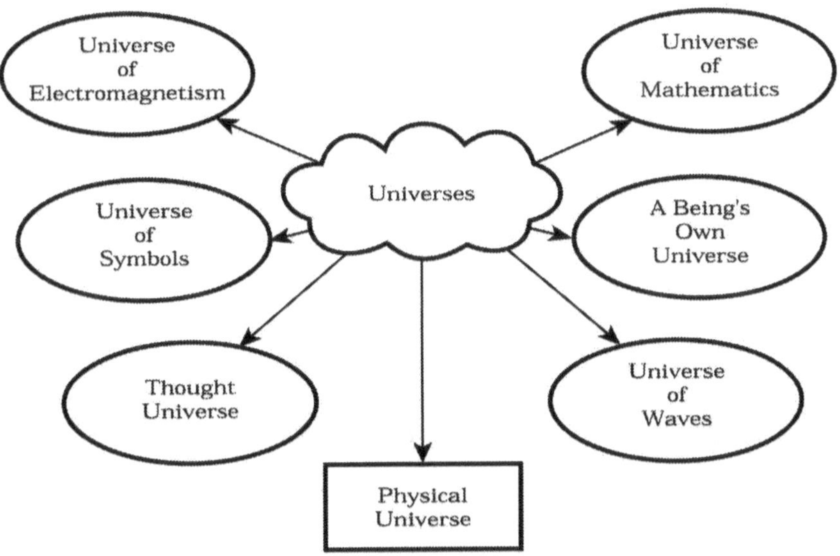

The Concept Of Universes

Even though all universes can be envisioned as one whole thing, they are actually constructed upon a series of actions, which are very precise with one following after another in an exact sequence. The sequence of actions is very important and must be understood thoroughly before one can gain a total comprehension of that universe.

"The reduction of the universe to a single being, the expansion of a single being even to God, this is love."

—Victor Hugo (1802-1885)
French Poet, Novelist and Dramatist

The Concept of Universe Generalized

By the term "universe," it is commonly understood that we mean the physical universe, which is a very specific universe, composed of one or more created spaces in which created masses and created energies are placed to interact so as to bring about mechanical time, more of which will be examined in great depth later.

The term "universe," as used in this book, denotes a generalization of the concept associated with the term "physical universe," that is to say, we have broadened (or generalized) the term "universe" to *mean any sphere of activity related by a common topic, that may or may not include all or a portion of the "physical universe."*

An example of a universe related to our study, other than the physical universe, is the universe of waves, which includes any and all waves existing in the universe. Components of such a universe would include mechanical waves, electrical waves, optical waves, bioelectric waves, thought waves, etc.

The interesting point about this universe is that it is a very wide sphere of activity and includes not only physical universe waves but also life waves (such as brain waves, telepathic waves, etc.) which are caused by the life force, and therefore, are not part and parcel or inherent in the physical universe. However, in this work, our primary focus is on the waves inherent in the physical universe.

"To confine our attention to terrestrial matters would be to limit the human spirit."

—Stephen Hawking (1942-)
English Physicist

How To Create Any Universe

We now examine the essential factors that enable us to construct any desired universe or a field of study such as physics or engineering in particular. By observation and the presented material in this book so far, we can conclude that to construct a universe we need to have the following series of actions:

a) **VIEWPOINT**—The first step is the assumption of a viewpoint. This is the first and foremost step in the creation of any universe.
b) **SPACE**—Have the viewpoint set up a region, which is delineated by a boundary surface. This region effectively creates a workable space in which all future creation can take place.
c) **PRIMARY POSTULATES**—Put forth a series of primary postulates, unconditionally true at all times, which are capable of creating different objects, each with its own distinctive characteristics and peculiarities. The term object, being the product of a postulate, should be taken conceptually and not literally, since in the field of arithmetic, for example, there are exact postulates, which have created "10 objects," numbers 0 through 9.
d) **WORKABLE & AUXILIARY POSTULATES**—Put forth a series of workable and auxiliary postulates, which define how the created objects will interact with each other. For example, in the field of arithmetic, there are four workable postulates, which create "four operations," which are addition, subtraction, multiplication, and division.
e) **LAWS, RULES & THEOREMS**—Obtain a series of laws, rules and theorems from these postulates, which will uniquely define the interactions and inter-relations

amongst the created objects. For example, in the field of arithmetic, the "commutative law" can be obtained purely by observation: 1+2=2+1. Therefore, steps (a-e) define all the considerations built into the field of arithmetic.

f) **APPLICATION MASS**—The final aspect of any postulated universe is the creation of its application mass, which approaches infinity in sheer number. For example, in the field of arithmetic, the application mass includes calculators, computers, computer codes, computer software and programs, ledgers, accounting balance sheets, abacus, etc.

The mechanics of creation of a universe as delineated in steps (a-f) is shown below.

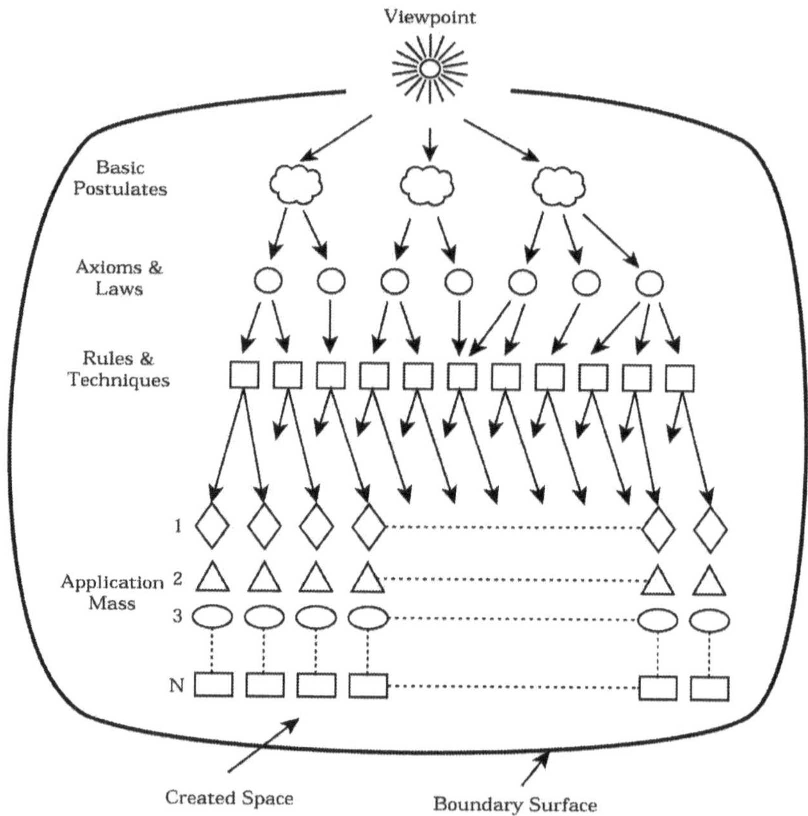

The Mechanics Of Creation Of Any Universe.

In summary, the first action after assuming a viewpoint and setting up a space, is to put forth a set of exact postulates (called primary postulates) to hold true at all times followed by a series of workable and auxiliary postulates and then define a series of laws, rules and theorems. The laws, rules, and theorems bring about and define an exact relationship between many things (such as the space and the postulates, etc.) and assist us in eventually arriving at the final destination of application mass.

The term "viewpoint" has a special meaning in this book and can be subdivided into a "postulating viewpoint" and an "observing viewpoint." In this section, when we say "viewpoint of a universe" we mean the former, which could also be referred to as the "postulator" or the "originator" of that universe. This term will be explored in more depth later.

"In creating, the only hard thing is to begin: a grass blade's no easier to make than an oak."

—James Russell Lowell (1819-1891)
American Poet, Critic, Editor and Diplomat

Our Seemingly Infinite Universe

As we look across the physical universe in which we live, we see a system of created particles and masses including solids, fluids, gases and plasma on a small scale; massive bodies including planets, meteors, stars, nebula, and asteroids forming the galaxies and clusters of galaxies on a larger scale, which span billions of light years and more.

With this general view of the physical universe, one is apt to assume an infiniteness about the physical universe, which is a total apparency and not an actuality. This apparency is caused by the seeming un-boundedness of the physical universe, which projects the idea that no matter where we go, we cannot get out of it.

This is only true if one has no road map out, then every movement would consolidate the concept of infiniteness.

We can visualize the concept of un-limitedness of something if we consider, for example, a fish swimming happily in an ocean. From the fish's viewpoint the ocean appears unlimited. This is so because the fish is interior to the space of the ocean. If the fish could exist exterior to the ocean, he would instantly realize the finiteness of the ocean and the whole matter would become an obvious apparency.

Thus, the key to the size of the physical universe lies in one's ability to get exterior to it, at which moment the physical universe becomes a finite universe. After years of consideration and much thought about this enigma, the puzzle can be considered resolved at this time, since pure observation of facts would bring vanishment to this apparency (see the Figure on next page).

"There must be a positive and negative in everything in the universe in order to complete a circuit or circle, without which there would be no activity, no motion."

—John McDonald (1916-1986)
American Writer

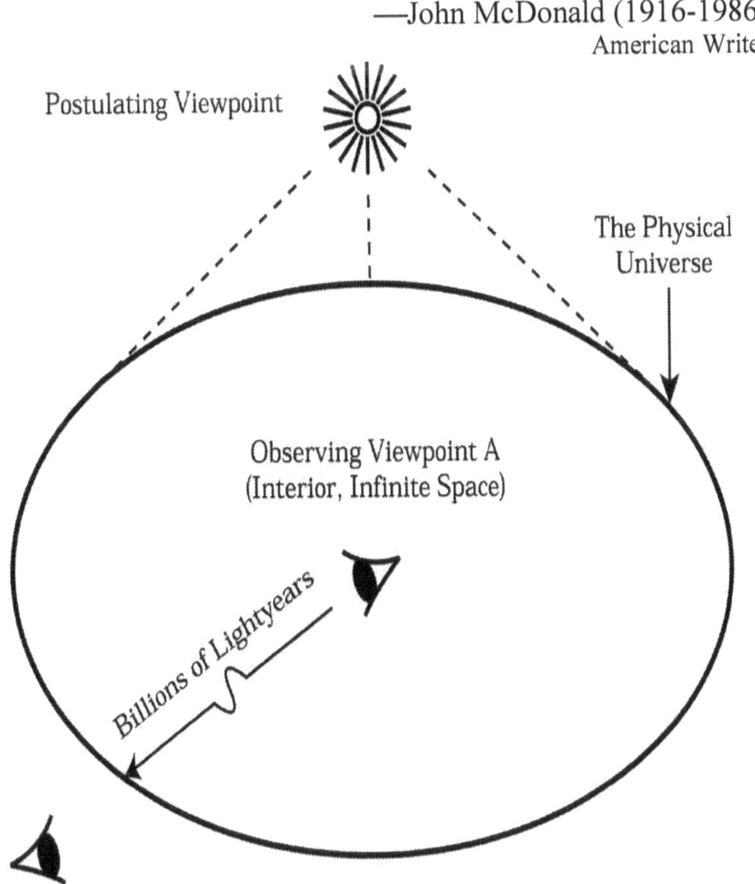

The Physical Universe
and its Viewpoint-Dependent Size

The Physical Universe And Its Relation To The Viewpoint.

PART II

The Essence of Physical Sciences

The Precious Knowledge

Since the beginning of recorded time, we seem to have been in a contest with the physical universe. When it comes to sciences, we seem to be totally inclined in the direction of trying to understand the physical universe with the idea of using the obtained knowledge for improving our own conditions of existence. This is done sometimes to a point of obsession. However, this obsession may be justifiable because our very survival on the physical plane depends on this point alone and no other.

"Science and art belong to the whole world, and before them vanish the barriers of nationality."

—Johann W. von Goethe (1749-1832)
German Poet and Novelist

In our daily existence, we are faced with enormous factors. We are faced with this unthinking and imposing thing called the physical universe, which cannot be dealt with or reasoned with on any level other than force. It only knows force. It is a universe of unintelligent force, which has submerged all living beings within it.

Being confined to a fragile and physical entity (called a body), which cannot tolerate much force, however, our main tool against the onslaught of the physical universe's demands and mandates is our organized bodies of knowledge called the physical sciences. This knowledge has been obtained through millennia of postulation and observation of the cause-effect of the observed phenomena in nature using the scientific methodology.

"Facts are stubborn things."

— John Adams (1797-1801)
Second US President

The knowledge about the physical universe has been handed down like a precious lore, improved and polished from generation to generation and has been preserved dearly. It is rightly so, because knowing this precious knowledge well and applying it effectively to ourselves and our surroundings has meant a higher level of survival for us, our progeny (children) and symbionts (codependents, pets, etc.).

"***Human history becomes more and more a race between education and catastrophe.***"

—H. G. Wells (1866 - 1946)
English Novelist and Historian

The Rise of Modern Sciences

"Reason, Observation, and Experience — the Holy Trinity of Science."

—Robert G. Ingersoll (1833-1899)
American Statesman and Orator

The rise of modern sciences ranks as one of the greatest events in the short history of Man in this corner of the universe. The fruits of modern sciences abound in terms of the many technological conveniences that it has created all around us. It has made life as a whole a more pleasant experience by relieving Man of more tedious work as well as raising our standard of living on many levels.

How did modern sciences evolve and how did they become so dominant in all aspects of our lives may be too long a story to be told in this book, but the beginning of how they achieved such a prominent position, may be of interest to us in this work.

About four and a half centuries ago in Europe, a number of brilliant minds started inquiring and inspecting the physical world in order to discover new truths and facts about it. The method they chose was a rational approach in examining the structure and natural forces of the physical universe and consisted of four major steps:

a) **Observation and imagination,**
b) **Generation of a working hypothesis, followed by**
c) **Experimentation to confirm the validity of the working hypothesis, and**
d) **Reiteration of this process until a valid hypothesis is obtained at which moment it becomes a principle.**

The new proposed method, as delineated in (a-d), started to make sciences as independent branches of learning and actually started a revolt against the dogmatic dictates based on ancient knowledge, which was handed down from the past, and was mostly obtained from religious and philosophical scholars, who were accepted as the final authority on matters of nature and the material universe. These dogmatic dictates had created a blind alley of investigation where any researcher in the field would be promptly dead-ended. This was a time period up to about the year 1500 A.D.

The modern science methodology to seek knowledge broke the traditional lore of learning by rote and consulting the past doctrines in favor of information, which was supported by experimental results. This pattern was about to set mankind on a totally different path of existence and produce much truth about our world that could benefit all mankind.

The beginnings of the scientific methodology and the bold revolt against ancient thinking are commemorated by two stellar names in the sciences:

I. **Galileo Galilei** (1564-1642), generally known as **Galileo**, an Italian physicist and astronomer, a founder of dynamics; and
II. **William Gilbert** (1544-1603), an English physicist, a founder of electric and magnetic sciences.

These were two names in the sixteenth century who signaled the dawn of a new age and thus will forever shine in the history of sciences. They were men of the Renaissance, the era of the great European rebirth in art, literature, and scientific thinking.

The importance of Galileo and Gilbert lies partly in their discoveries but more importantly in the methodology they presented for scientific investigation. They proposed a method of *observation, hypothesis, and controlled experimentation to verify the information and obtain the unknown from the known.* They advocated *measurement and analysis, rather than the unequivocal acceptance of any presented statement, which was the existing scholastic tradition of the day.*

Both men (i.e., Galileo and Gilbert), by persistent investigation of natural laws, laid the groundwork for modern experimental sciences, thus creating the towering giant of sciences that is today. Moreover, they set the ground rules for pragmatic procedures that started a revolution in scientific thinking and methodology. Their defying of the ancient dogma and the scholar authorities of the day, declared a shift from the philosophical/religious age to the beginning of the scientific era. For this, both were denounced and eternally anathematized (i.e., cursed or consigned to damnation) and thus attained immortality in a peculiar way!

By a broad and sweeping look across the morass of life from the ancient ages to the present time, one can see that sciences are the main proponents of truth in any particular subject, since they are able to provide a series of exact principles, which leads to very exact and repeatable results.

"Science does not know its debt to imagination."

—Ralph Waldo Emerson (1803-1882)
American Essayist

Today, the hard sciences are waging war against superstition and ignorance, which are the main causes for resorting to authoritative sources of information in the first place. Even though it may lead to a solution—a temporary remedy on a short term basis when one is faced with a problem, the final downfall of anyone using arbitrary solutions containing arbitrary factors is just a matter of time. The resulting solution becomes unworkable on a long term basis because the problem was not solved in a unique manner as prescribed by the uniqueness axiom, of which more later.

Thus, it can be clearly observed that with a display of bravery as set forth by these two prominent scientists, the door to scientific exploration was firmly swung wide open. Over a period of four and a half centuries of scientific discoveries and revelations, Man was enabled to soar to greater heights than ever imagined possible and has recently emerged as the victorious ruler of earth. He currently stands on the verge of space exploration and space travel, to conquer the rest of this enigmatic universe.

"There are in fact two things, science and opinion; the former begets knowledge, the latter ignorance."

—Hippocrates (460 – 377 BC)
Greek Physician, known as the Father of Medicine

The Theory of Relativity

The Theory of relativity is a theory based on mechanical observation of the same phenomenon from different frames of reference having different velocities and accelerations. In his famous work, Einstein proposed a theory of physics, which recognizes the universal character of the propagation and speed of light and the fact that the measurement of time, space, etc., depends on the speed of the observer who is performing the measurement.

"Relativity teaches us the connection between the different descriptions of one and the same reality."

—Albert Einstein (1879-1955)
German Born American Physicist, Nobel Prize in 1921

It has two main divisions:

a) The Special Theory of Relativity (circa 1905), and
b) The General Theory of Relativity (circa 1915).

I. Special Theory of Relativity proposes that all natural laws are the same for observers in "inertial reference frames," which are frames of reference moving at constant speed.

Inertial reference frames contain no gravitational fields. Such a frame of reference describes a created space, which is linear.

According to "special theory of relativity" we can see that the laws of physics are invariant to a transformation from one reference system to another moving with a linear, uniform, relative velocity.

In other words, the course of natural phenomena is unaffected by a non-accelerated motion of the coordinate systems to which they are referred, and therefore, all reference systems moving linearly and uniformly relative to each other are equivalent.

For example, all the laws of physics on every location on earth (e.g., a point at North pole and another one at equator, where they are traveling at different angular speeds) would provide the same answer and will be preserved in form because the reference frames that the same two events are measured in, are all traveling at the same linear and uniform speed.

However, the same two experiments: one on earth and another at location "x," on planet "y" in the star system "z," if performed accurately would not necessarily provide the same answer because of different planetary relative accelerations as well as gravitational fields, where the reference frames are located.

The second part of Einstein's special theory of relativity is more remarkable and states that "the velocity of propagation of an electromagnetic disturbance in free space is a universal constant "c," which is independent of any reference system."

This concept is quite contrary to our experience with mechanical or acoustical waves in a material medium, where the velocity is known to depend on the motion of the observer relative to the propagation medium. Many experiments have been devised to examine the validity of such a radical assumption; however, the results have all indicated that this concept is consistent with all of the known electromagnetic and optical observed phenomena.

It should be pointed out that the concept of the universal constant "c" is valid at low speeds of relative travel. However, it is highly unlikely that such a proposition would hold valid if the observer's reference frame travels at or close to the speed of light. Since such high speeds of travel for reference frames are impractical and unachievable at this time; therefore, this concept remains valid at low speeds. However, future work at high speeds may modify it.

"Men love to wonder and that is the seed of our science."

—Ralph Waldo Emerson (1803-1882)
American Essayist

II. The General Theory of Relativity generalizes the "Special Theory of Relativity" to nonlinear and non-inertial frames of reference, where gravitation is incorporated and in which events take place in a curved or nonlinear space.

An example of such a curved space would be travel of light on a curved path (i.e., not in a straight line) in strong gravitational fields. Expanding on this example, we can see that the light traveling from the sun to the earth would also get bent as shown below.

The path of light is not a straight line but actually bends due to sun's strong gravitational field as it takes off from the sun. The light bends once again as it enters earth's gravitational field.

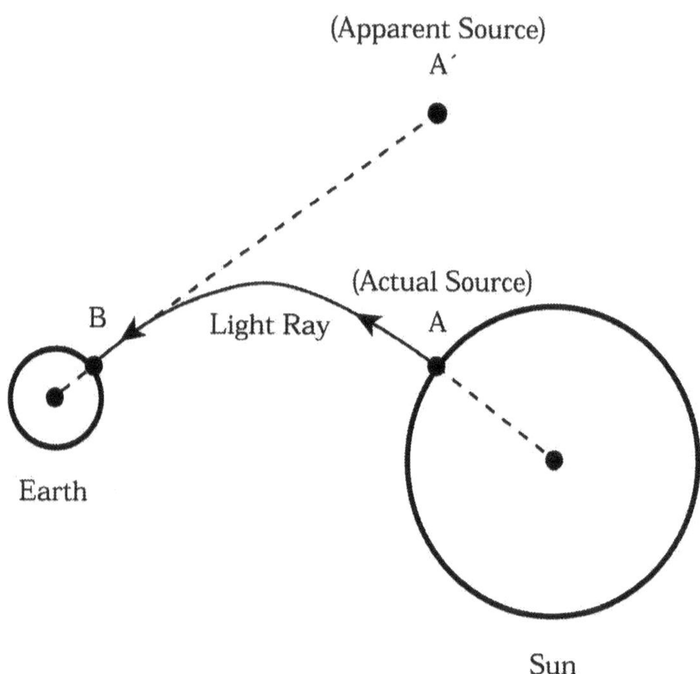

Note: Diagram is not drawn to scale.

Bending Of Light Due To Gravitational Fields Of Sun And Earth.

The important thing to remember is that this curved motion of light is not the property of space but of the force field that "colors" that region of space momentarily. This curved traveling path is due to strong gravitational lines of force. The created space remains linear regardless of the energy fields in it.

In short, we can conclude that what Einstein was attempting to put forth, in a round-about and yet scientific way, was the fact that there is no such thing as an absolute data, when dealing with materials connected with the physical universe.

In other words, it all depends on the frame of reference in which one is conducting the measurement. The data so obtained is all relatively true depending on the point of view of the observer.

"A man sits with a pretty girl for an hour, it seems like a minute. He sits on a hot stove for a minute, it's longer than any hour. That is relativity."

—Albert Einstein (1879-1955)
German Born American Physicist, Nobel Prize in 1921

Absolutes

The material components that make up the physical universe are caught between two absolutes: zero and infinity. Zero and infinity are two examples of an "***Absolute***," which is an important term and is defined as: **a)** *That which is without reference to anything else and thus not comparative or dependent upon external conditions for its existence (opposed to relative);* ***b)*** *That which is free from any limitations or restrictions and is thus unconditionally true at all times.*

By pure observation, we can state that the material components that make up the physical universe are caught between two absolutes: zero and infinity.

To understand this concept, let us consider one of its components such as temperature, which is directly associated with energy. It is a known fact in physics that a temperature of zero on the absolute scale, cannot be reached but asymptotically approached. This is because a temperature of zero degrees Kelvin (on the absolute temperature scale) designates a state of "no energy" or "no motion" i.e. a "complete static," which violates the make-up of the physical universe (a universe purely based upon motion) but can be considered as a mathematical absolute.

Moreover, it is also an obvious fact that a temperature of infinity, which is a state of infinite energy, can not be reached. These simple observations bring us to the conclusion that the energy in the physical universe is bounded between two absolute energy levels: zero and infinity.

Similarly, we can take another component, such as matter, and reach the same conclusions. To make the proper conclusions for this component, we can use the equivalence of matter and energy and the fact that matter is a condensation of energy, and therefore based on the same reasoning set forth for energy, a zero or infinity of matter is also impossible to attain.

For the last real component, i.e., space, we can observe that any portion of the space in the physical universe is filled with either matter or energy, therefore a zero or infinity of space cannot be achieved; since such a condition leads to a zero or infinity of energy which is impossible to attain as far as the physical universe is concerned.

It should be noted that an absolute vacuum of space (i.e. space void of matter or energy) is impossible to attain since, even in outer space where there is no air, all types of particles such as space dust, photons, cosmic rays, etc. still are contained in every region of space under consideration.

We can observe that the main components of the physical universe are matter and energy (both finite), existing in a linear space (finite), and constantly interacting with each other leading to time. Thus time on any scale of measurement becomes finite since it has a definite beginning and a finite end. Thus we can conclude that the physical universe consists of three tangible and real components (matter, energy and space) and a fourth intangible and pseudo-component (time). In other words we are dealing with a universe where each of its components can never achieve absolutes.

We can extrapolate this concept of "non-absoluteness," in a deductive logic fashion, to any other physical entities, particularly electrical quantities such as power, current, electric field, etc., as well as any other related physical laws, facts, etc. This observation connotes that that all considerations derived from or concerning the physical universe has a finite value and do not achieve zero or infinity in importance or validity, which means that they only have relative importance.

Furthermore, because all physical quantities have a value relative to one another, then they can be plotted on a graduated scale of importance or seniority, extending from zero to infinity, but not touching zero or infinity since these two are absolutes. Using this graduated scale of plotted data, we can now see that the concept of "relativeness" can be generalized and applied not only to physical quantities but also to abstract qualities as shown in the Figure below.

The Graduated Scale Of Abstract Concepts Or Physical Quantities From The Lowest To The Highest Absolutes..

Furthermore, since there is no absolute datum (i.e., a datum that has infinite importance or is true unconditionally at all times), then the only absolute considerations that can be designated as such, is perforce by means of *an agreement, or more precisely a postulate*.

The idea of using a graduated or gradient scale is not a new concept in mathematics but using it to plot relative data is! It is a common practice to plot for example real positive numbers on a scale that extends from zero to infinity. However, what we are doing here is actually generalizing this gradient concept in mathematics, which up to now has only been applied to numbers. Using it to plot out any and all data (numerical, conceptual, physical, etc.) concerning the physical universe materials (e.g. voltage, current, power, etc.), is a way to plot out their order of importance, validity, etc.

Since this graduated scale is based on the data derived from the finiteness of the physical universe, it would not permit any of the absolutes at either end (0 or ∞), to be achieved but only approached.

"Science can purify religion from error and superstition. Religion can purify science from idolatry and false absolutes."

—Pope John Paul II (1920-2005)
The Polish Pope

The Uncertainty Principle

This is an all-pervading principle, which is interwoven in the woof and warp of the physical universe on a macroscopic or microscopic level. The uncertainty principle is an important operating principle in the microscopic universe and is quite revealing of a number of subatomic phenomena.

The Uncertainty principle has had serious ramifications, particularly in the area of Quantum physics but hardly poses any significant measurement error at the Classical Physics level of observation.

"In the middle of difficulty lies opportunity."

—Albert Einstein (1879-1955)
German Born American Physicist, Nobel Prize in 1921

The Uncertainty Principle, also called the "Heisenberg Uncertainty Principle" or the "Indeterminacy Principle," *states that the accurate measurement of an observable quantity (such as position or time) necessarily produces uncertainties in one's knowledge of the values of other associated observables (such as speed or energy).*

It can be stated, in very simple terms, that the principle of uncertainty is actually a statement about the dichotomy of a "known and unknown" pair. In other words it states that every microscopic particle (e.g., an electron, a photon, etc.) can be associated with an accurately known value regarding one of its characteristics (e.g. speed, etc.) whereas all other measurable characteristics (e.g. energy, position, etc.) are unknowable but become knowable only through a certain range of tolerance, i.e., an error is introduced in their values.

This concept applies particularly to a moving object's speed and location. If one can accurately determine the location of the object, then its speed cannot be accurately established. Similarly, if one establishes the energy of a moving object with a high degree of accuracy, then its time of existence at a certain location cannot be accurately determined.

"I tore myself away from the safe comfort of certainties through my love for truth – and truth rewarded me."

—Simone de Beauvoir (1908-1986)
French Writer and Novelist

The concept of uncertainty, as innate as it may seem to the microscopic universe (wherein quantum physics is but a subset), has several reasons for its existence, amongst many the following are of significance:

a) We are dealing with extremely fast moving particles in the microscopic universe, which are not as easily measurable compared with their macroscopic counterparts.
b) Particles at the microscopic level show their true nature, which is a condensed energy form with their exact location quite unclear but only to be determined by a probability value.
c) Microscopic particles are more like a wave packet, which is considered to be an ultra-small standing wave having one dominant peak and several smaller peaks and valleys. Particles at this level of observation are thus, seen to be not composed of one coherent solid form but rather made up of a close-woven, alternating energy form.

"What men want is not knowledge, but certainty."

— Bertrand Russell (1872-1970)
English Logician and Philosopher

The Uniqueness Axiom

The realm of "*Technical Solutions To Problems*" as a whole class of data can be analyzed by knowing its most central datum, "The Uniqueness Axiom."

The Uniqueness Axiom is an often used but seldom known axiom in engineering, which can be defined as: *Given an exact set of necessary and sufficient initial conditions or boundary conditions, any problem in the physical universe has a unique and exact solution.*

The important point to remember here is that the Uniqueness Axiom only applies when the problem under consideration exists in the "*physical universe*" with all necessary assumptions exactly specified.

This is unlike the social sciences where there may be many arbitrary solutions to one problem; all seem to provide a seemingly satisfactory answer to a social problem, but none of the solutions are unique (i.e., all are arbitrary solutions!).

This is where modern science and engineering fields, with the use of uniqueness axiom, have diverged greatly from social sciences and have assumed a commanding presence in our today's society and the workaday world.

"A work of art is the unique result of a unique temperament."

—Oscar Wilde (1854-1900)
Irish Poet, Novelist, and Dramatist

Relativity of Knowledge

We can extrapolate and expand Einstein's theory and obtain a highly generalized principle: *The Principle of Relativity of Knowledge*. This generalized version of the theory of relativity is vastly workable, since it clearly demonstrates to one the highly viewpoint-dependent nature of the world in which we live. This principle goes far beyond the perimeter of scientific arena and actually applies to any and all parts of our existence.

Utilizing the concept of "Absolutes" in the physical universe, we are now ready to define an important principle, which has ramifications throughout the entire field of science and technology. That principle, of course, is "***The Principle of Relativity of Knowledge***," which is defined as:

"Any datum, such as a natural law, a principle, a fact, a measured value, etc., (which is derivable from the postulates and axioms), can never have an absolute value, but only a relative value whose relationship to other data is based on the point of view of the observer."

This principle actually broadens the concept of relativity and brings out a much bigger arena for examination than has ever been envisioned.

" I sometimes ask myself how it came about that I was the one to develop the theory of relativity. The reason, I think, is that a normal adult never stops to think about problems of space and time. These are things which he has thought about as a child."

—Albert Einstein (1879-1955)
German Born American Physicist, Nobel Prize in 1921

Applying this principle to a field of study such as “electricity and magnetism,” at once demonstrates to us that the concepts therein are only relatively true and lie somewhere on a gradient scale and therefore, should never be considered to be absolutes!

For example, let's consider one of the physical sciences (e.g. Physics) and ask ourselves this question, "Can Physics contain an absolute law, which would be true under all conditions?"

Using the concept of "Absolutes," we can see that such a condition of absoluteness is impossible since all laws are derived from postulates. We know that postulates of a science are the only absolutes and anything derived from a postulate is perforce conditional or relative and therefore not absolute. This applies to all of our beloved laws of physics, simple or complex, presented in modern textbooks! They all have definite limitations and thus are relatively true.

This is a direct consequence of working within a finite physical universe. Even if one achieves a workable law, which is expressed and cast into absolute terms of an exact mathematical equation, however, when that equation is put into practice, it immediately falls into the category of a relative law and can never be held true unconditionally in all locations and under all conditions without any further qualifications. This applies to all laws such as Newton’s Laws, Maxwell’s Equations, etc.

From the above discussion and example, one can conclude that, "The only absolutes in a science are the postulates. All of the physical laws are relatively true and can never achieve absolutes in actual practice, unconditionally and for all times. Absolutes can only be approached."

"***Mathematics would certainly have not come into existence if one had known from the beginning that there was in nature no exactly straight line, no actual circle, no absolute magnitude.***"

— Friedrich Nietzsche (1844-1900)
German classical Scholar, and Philosopher

Moreover, we can observe that all of the scientific laws do not have a monotone order of importance. Each law has a different weight, which should be evaluated properly relative to other laws before it is applied to a practical situation therein lies their inherent relative value.

"What wisdom can you find that is greater than kindness?"

Jean-Jacques Rousseau (1712-1778)
French Philosopher and Writer,
Whose Novels Inspired the Leaders of the French Revolution

An important **CAVEAT** is worth mentioning here *and that is, "Natural Laws" when cast into exact mathematical forms can appear to be absolutes. It is only an apparency, because the mathematics that is used to represent these laws can only express them in absolute forms. Thus, mathematical equations and their corresponding exact solutions serve only as guidelines in practical applications— to find absolute solutions to idealized problems!*

"I have just three things to teach: simplicity, patience, compassion. These three are your greatest treasures."

—Lao Tzu (600 -531 BC)
A Chinese Philosopher

The Pyramid of Knowledge

Workable knowledge is like a pyramid, where from a handful of common denominators efficiently expressed by a series of basic postulates, axioms and natural laws, which form the foundation of a science, an almost innumerable number of devices, circuits, and systems can be thought up and developed. The plethora of the mass of devices, circuits, and systems generated is known as the "application mass," which practically approaches infinity in sheer number.

"The art and science of asking questions is the source of all knowledge." —Thomas Berger (1924-)
American Novelist

By "**Application Mass**" we mean, *"A very specific type of mass consisting of all of the related masses that are connected and/or obtained as a result of the application of a science."* This includes all physical devices, machines, experimental setups, and other physical materials that are directly or indirectly derived from and are a result of the application.

The inter-relation between the postulates and the application mass of a science is an important point to grasp, since the foundation portion never changes (a static) while the base area of the pyramid is an ever-changing and ever-evolving arena (a kinetic), where this evolution is in terms of novel implementation techniques and new technologies.

It can be observed that the fundamental postulates of a science, as a whole, form the bedrock upon which all natural laws rest. Furthermore, we can see that the postulates and the discovered natural laws, altogether, form the foundation of a science. It is an important concept, which is omitted in the majority of scientific texts. All the remaining considerations such as scientific conclusions, technical data, design methods, rules, etc., as well as their byproducts (space, energy, matter, and time) and the entire application mass of the subject is derived from the foundation.

"Kindness in words creates confidence. Kindness in thinking creates profoundness. Kindness in giving creates love."

—Lao Tzu (600 -531 BC)
A Chinese Philosopher

The Figure below shows the pyramid of knowledge in any science.

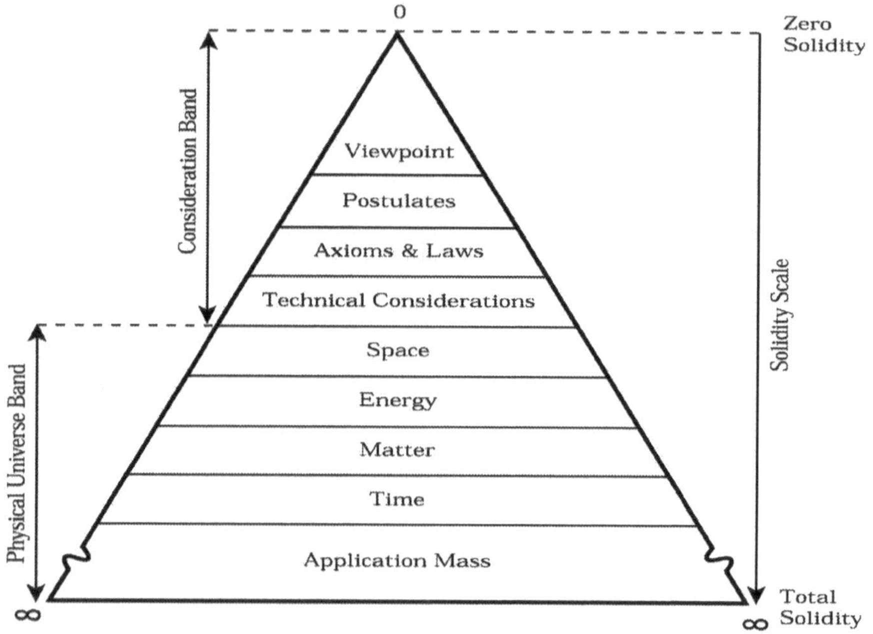

The Pyramid Of Knowledge In A Science.

In simpler terms, ***the closer the laws of a science are to simplicity and the basic postulates, the higher the workability of the science.***

Expanding this pyramid for more exactness and details, we obtain the pyramid of knowledge in a workable science (such as engineering) as shown on the next page.

"Only the educated are free."

—Epictetus (55 – 135 AD)
Greek Philosopher

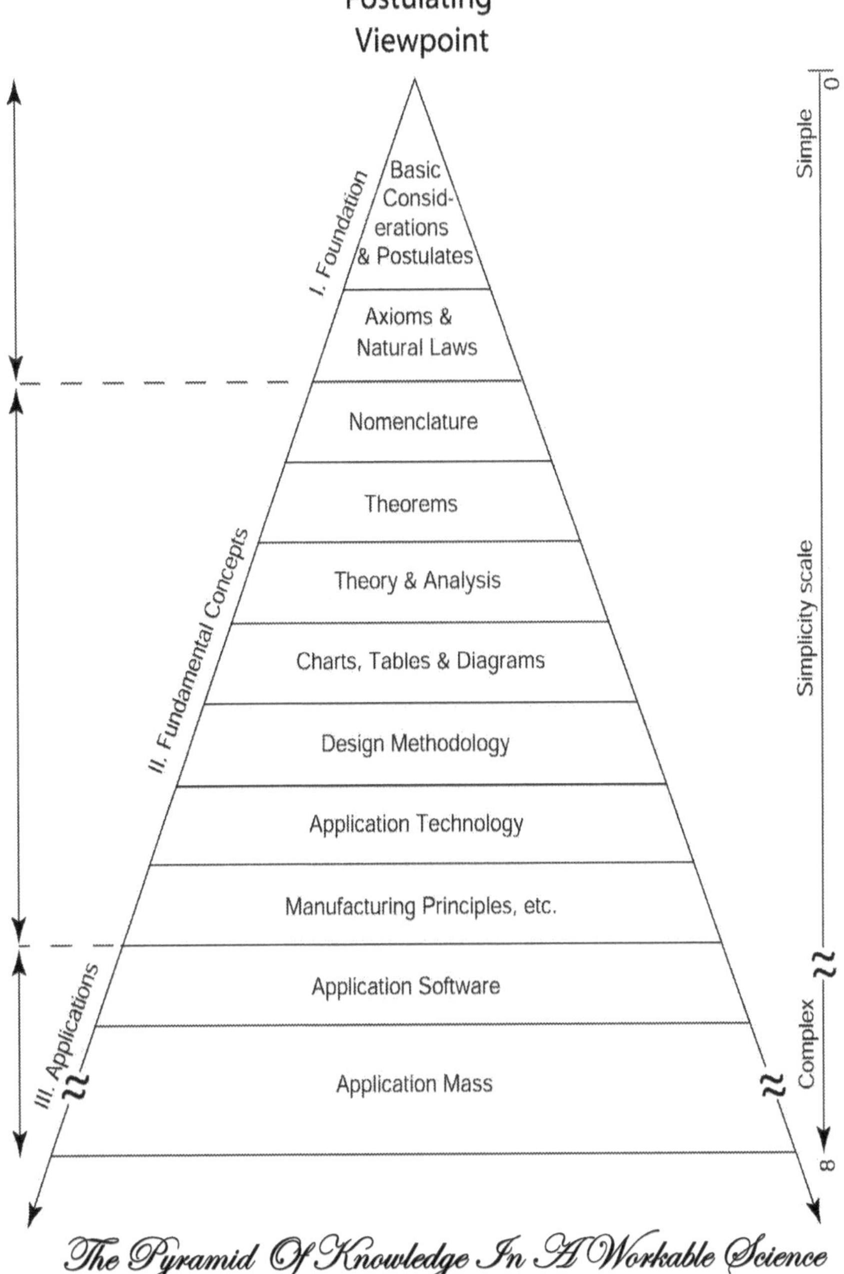

The Pyramid Of Knowledge In A Workable Science (Such As Engineering).

The Circle of Knowledge

By observation we can generalize the pyramid of knowledge to include all of the extant sciences. This generalization would take the shape of a circle or a pie as shown on the next page.

From this Figure we can see the entire field of knowledge of mankind about the physical universe takes on the form of a circle (or a pie), where each workable science is a slice of the pie.

At the center of the circle lies the viewpoint (or a collective viewpoint), which puts forth a series of the fundamental postulates concerning the physical universe. This factor is held in common by all of the sciences.

"Science is built up of facts, as a house is built of stones; but an accumulation of facts is no more a science than a heap of stones is a house."

— Jules Henri Poincaré (1854-1912)
French Mathematician, Physicist, and Philosopher of Science

This common area, at the center of the circle, forms an inter-relatedness amongst all sciences. We will define and elaborate on the more significant aspects of this pyramid in the forthcoming sections.

"***It is the mark of an educated mind to be able to entertain a thought without accepting it.***"

—Aristotle (384 – 322 BC)
Greek Philosopher, Scientist and Physician

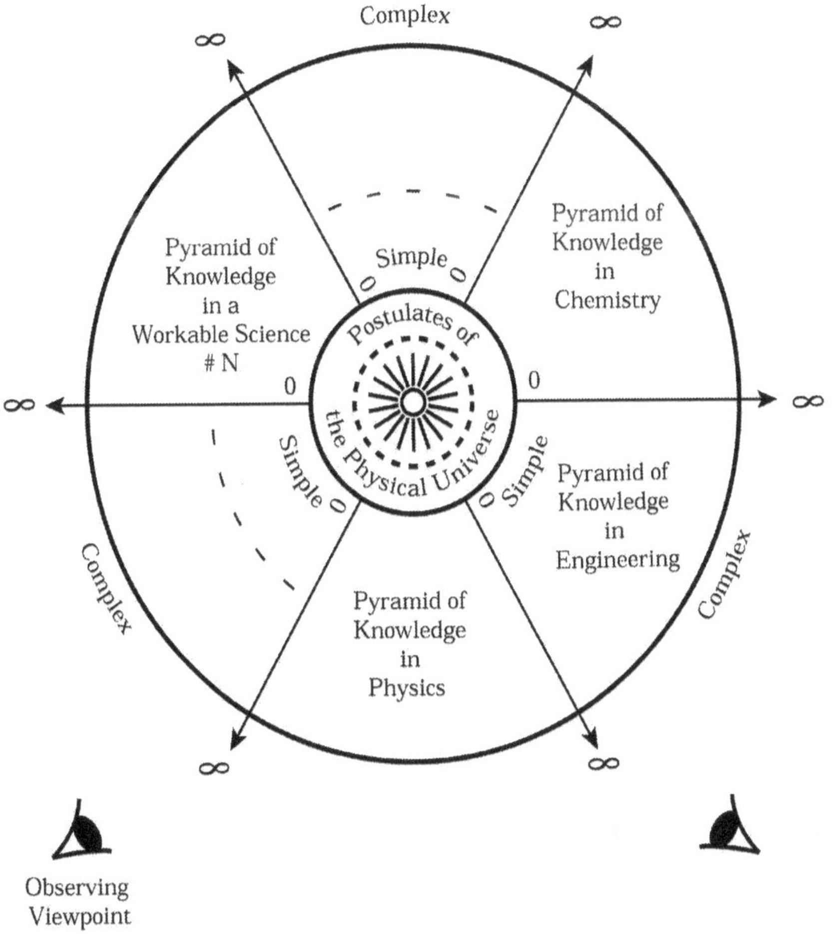

Complex

Pyramid of Knowledge in a Workable Science # N

Pyramid of Knowledge in Chemistry

Pyramid of Knowledge in Engineering

Pyramid of Knowledge in Physics

Postulates of the Physical Universe

Simple

Simple

Simple

Complex

Complex

Observing Viewpoint

The Circle Of Knowledge And Its Relation To The Pyramid Of Knowledge.

"It is not enough to have knowledge, one must also apply it. It is not enough to have wishes, one must also accomplish."

Johann Wolfgang von Goethe (1749-1832)
German Playwright, Poet, Novelist and Dramatist

How the Universe Is Explored in Sciences

It should be noted that the concept of the physical universe as wide as it may be in general, becomes a cut-down version of the actuality when it is taken up in the physical sciences, which are very limited and narrow arenas of study. When we say the "the physical universe in sciences" we actually mean *"that portion of the world (in terms of masses, spaces, energies and time frames), which a science directly influences."*

"Science progresses best when observations force us to alter our preconceptions."

—Vera Rubin (1928-)
American Astronomer

For example, when we say "physical universe" in the field of electricity and magnetism, we actually mean electric currents and voltages, which run in the created spaces of wires and cables, transistors, electronic parts, Integrated Circuits (Ics), etc.

Using the physical universe in this manner, the concept of the creation of a universe would no longer be a general statement, rather limited to the narrower arena of electricity and magnetism.

By doing this, the process of creation of a universe becomes more meaningful and comes to life and thus no longer appears as an abstract concept. This is an important point, which should be kept in focus throughout this book.

"***There are no such things as applied sciences, only applications of science.***"

—Louis Pasteur (1822 – 1895)
French Chemist and Microbiologist

Technical Nomenclature

Any scientific subject, when communicated verbally or in writing, is put into a language format, which at once makes it quantized. In other words, a concept as fluid and as continuous it may be in one's mind, the moment it is presented for communication to others has to be transformed into a language form, which is made up of information quanta called "words." This transformation process essentially makes any "scientific concept" no longer continuous but rather quantized or pulsed (borrowing a term from electrical signals nomenclature).

The basic building blocks (or quanta) of information of any science, are its technical terms or terminology. Each technical term (or quantum of information) carries with it an exact package of information, which is embodied in its definition. Therefore, in examining any scientific subject, the first point of encounter is with its specialized words, which are a series of precisely defined technical terms. The collective sum of these specialized words is commonly referred to as "nomenclature."

The full mental grasp of each of these specialized words by one who is traveling on the road to knowledge is an important milestone in one's progress and ascent to higher levels of realization and understanding.

Needless to say, that neglect to understand or misunderstand these essential terms could wreak havoc in one's mind and prove to be disastrous in one's progress along the road to knowledge. Some of the consequences of such a neglect and/or misunderstanding is loss of interest, focus, and a general state of dislike and confusion leading to gross misapplications.

The important point to grasp here is that the pioneers and founders of a scientific field of study felt that it was necessary to invent these terms, and they, themselves, communicated their discoveries, laws and principles using these same very terms.

Now, years later, a student of science comes along and tries to learn the subject by reading about the laws and principles of the subject, which the pioneers have earlier discovered and written about with tremendous clarity. "Without the student's full comprehension and mastery of the specialized words and technical terms, what are the odds of a successful communication with any given subject?" one may surmise. The obvious answer is nil.

For any student of science to come along and try to understand the same very discovered laws and principles with a complete disregard for the terms, is extremely adventurous, to say the least! As obvious as this observation may seem to many, yet there are a great many students who have no traffic with a technical dictionary while studying a technical subject. It can be seen that they are inviting nothing but a great amount of complexity of thought and perplexity of application into their studies, unknowingly.

"If you think in terms of a year, plant a seed; if in terms of ten years, plant trees; if in terms of 100 years, teach the people."

—Confucius (551-479 BC)
China's Most Famous Teacher, Philosopher, and Political Theorist

Furthermore, it is a well-known fact that having an inadequate comprehension of the terminology is one of the leading causes of confusion and misunderstanding of a subject. Therefore, the terminology of a science forms an important part of the subject, and mastery of any subject requires mastery of its terminology along with its accurate definitions. This means that a student, when faced with a stream of information regarding a scientific concept, needs to fully grasp each quantum of information (i.e. each technical term), with the exact definition that was originally intended by its author, before the whole scientific concept is grasped.

So the way one becomes successful in the study of a science is through the full comprehension of its nomenclature. In other words, one enters through the gateway of a science through the comprehension of its very first term and continues on with full attention on the definitions of its nomenclature throughout the course of study, since there is no other road to a higher plateau of knowledge or critical thinking.

Therefore, we can see that one successfully enters the gate of a science through the comprehension of its language, which is made up of quanta of information (i.e., scientific terminology). We can actually express this observation by the following conclusion:

In any scientific field of study, the comprehension of specific and exact definitions of relevant terms are essential if one intends to apply the subject successfully and/or communicate the postulates, principles, laws, observations, problems and solutions to others.

The figure below shows how concepts are coded and condensed into terms and symbols for ease of communication and how, in order to obtain the meaning of the terms and symbols, one must expand them mentally for proper comprehension of the intended concepts.

Knowing well that terminology plays a superior role in one's comprehension of a subject, precautionary measures have been taken throughout this work to pay special attention to new terms by defining them as they are introduced. Additionally and for easy reference, a relatively comprehensive glossary of important technical terms is provided at the end of the book.

"Learning without thought is labor lost."

—Confucius (551-479 BC)
China's Most Famous Teacher, Philosopher, and Political Theorist

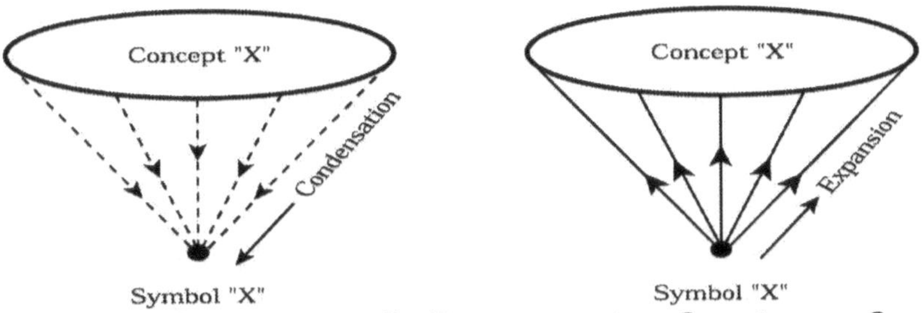

Concept "X"
Condensation
Symbol "X"

Concept "X"
Expansion
Symbol "X"

The Condensation Of A Concept Into A Symbol And Expansion Of A Symbol Into A Concept In One's Mind.

⊶ ✶ ❁ ✶ ⊶

Solutions in Sciences

Modern human beings started the process of survival on a physical plane on this very far and removed planet some 35,000 to 100,000 years ago. Ever since that time we, as a species, have relentlessly attempted to solve the myriads of problems that life has handed us on a daily basis. Our stock-in-trade has been our power of reasoning and observation far above other creatures'.

Now in our modern age, we have inherited a hectic world from our ancestors, a world which has us surrounded on a constant basis by a myriad of present and future survival problems. Our modern world, which has emerged as highly technical, has an extreme propensity toward correct technical solutions. This is where engineering and physical sciences march in and attempt to help us in solving our survival problems in a methodical fashion.

With the advent of electricity in its many forms (voltage, current, RF waves, light, etc.), it can be said with a great degree of certainty that the game, as far as the conquest of the physical universe is concerned, has been over for a number of years.

"The important thing in science is not so much to obtain new facts as to discover new ways of thinking about them."

—William Bragg, Sr. (1862-1942)
British Physicist

The success of human beings, in the business of survival accompanied by a tremendous onslaught of automated machinery, has increased the population of our species. However, we are ill-prepared to handle the social aspects of our increased population and the enormous advances in the field of physical sciences which would cause dangerous implications in the wrong hands.

The lack of development of a comparable and genuine social science (free of biased ideologies) that can handle human beings as a species, combined with the lack of a workable technology to handle individuals, groups, nations and earth as a whole on an axiomatic level and a unique-solution type basis, is taking our current society and as a result, the earth in a totally uncharted territory, which can be humorously referred to as "Repeating History."

As a species, we have no competition on earth, and with the game of survival on a physical plane already won over all other creatures of earth, the only game left is bettering ourselves spiritually and properly managing groups of our own kind, which are increasing in number and size at a rapid rate, by making use of the enormous untapped potential available on the mental plane.

"When the solution is simple, God is answering."

—Albert Einstein (1879-1955)
German Born American Physicist, Nobel Prize in 1921

Solutions: Unique Vs. Arbitrary

Any unique and exact solution is surrounded by an almost infinite number of "arbitrary solutions" which are all deviations from the central datum. These deviations are all based on an introduction of an "arbitrary factor" at some point along the way in the solution process. Thus, we can conclude that:

Use of arbitrary factors produces arbitrary solutions, which are all deviations from the exact and unique solution of any given physical universe problem (see the Figure shown below).

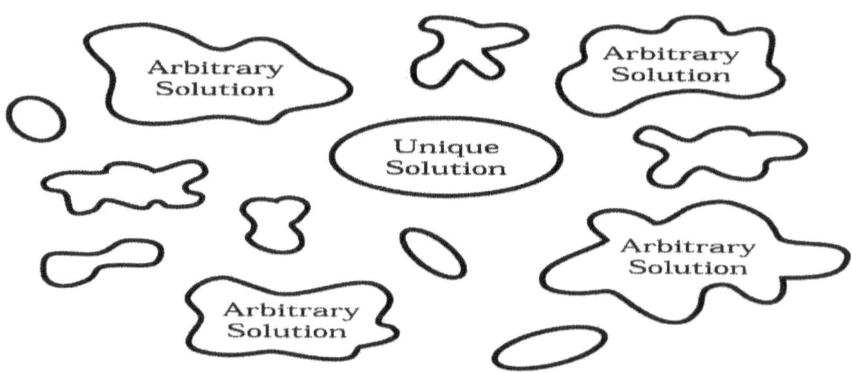

The Unique Solution In A Sea Of Arbitrary Solutions.

It is important to note that the uniqueness axiom only applies to "analysis" type problems and excludes all "design" type problems. This is a rather obvious observation, since for example, while there is only one unique answer to a broken TV set (an exact part number!), a skilled designer could easily design several electronic circuits, leading to different TV models, all functioning satisfactorily.

It can be further observed that the introduction of an arbitrary factor generated by authoritative sources or based on one's misunderstandings will invite the introduction of more arbitrary factors. This is where the old maxim "Deviations from truth of a situation beget more deviations," meaning, "Lies beget more lies," comes to mind and illuminates this concept even further. Moreover, because the concept of "arbitrary factors" applies to a much wider sphere of existence on a physical plane; therefore, it is applicable to the field of science and engineering as a subset of the physical universe.

We can have a reverse look at the "number of arbitrary factors in a subject" and by observation conclude that those fields of study (e.g., arts, politics, humanities, etc.), which mostly depend on authoritative opinions for their source of data, contain the lowest number of natural laws.

An example could illustrate this point further. Let's see how the concept of "unique and arbitrary solutions" apply to a field of study such as "politics" and what conclusion can we make about this subject.

To answer such a question, we need to observe that the field of politics has been pockmarked with arbitrary ideas of politicians, opinion leaders, political educators, senators, newspaper editorials, etc. to a point that a completely unworkable subject has been created.

This is a field of study, where lives of human beings, their happiness and future survival on a national or a global level hang in the balance and yet nothing is known about it on an inexorable and axiomatic level.

Nevertheless, the field of politics has been aggrandized and is now called "political science," which is an oxymoron of terms to the nth degree. It is a field of study where many strange and arbitrary solutions to global or societal problems are constantly offered to the populace and passed as workable solutions. The public instinctively senses this arbitrariness but cannot identify it as such in terms of axioms and natural laws, thus, the political charade goes on.

There is not one single axiom or natural law that has been discovered or stated in any textbook published on politics and yet it is one of the most vital fields of study to mankind, far more important than physical sciences.

It has not achieved the status of "unique solution type of science" as all physical sciences have. Thus, it has stagnated for centuries, if not millennia, as a field of study.

Furthermore, we can conclude that the solutions to the problems of mankind as proposed by politician "X" are totally arbitrary and not unique, mostly unworkable on a larger scale, even if somewhat workable on a micro-scale and "band-aid" level of operation.

Generalizing the concepts discussed in the example above, we can make a general conclusion: Any field of study (such as politics, economy, arts, etc.), being filled with many arbitrary factors and authoritative opinions, which form the foundation material of that subject, shall perforce become unworkable.

"The whole history of science has been the gradual realization that events do not happen in an arbitrary manner, but that they reflect a certain underlying order, which may or may not be divinely inspired."

— Stephen Hawking (1942-)
English Physicist

Solving Problems

To solve any technical problem uniquely and not arbitrarily, any competent scientist would start by breaking down the problem into areas of similar magnitude and getting a solution for each area and then integrating them back together for the final solution. The technique is very simple and can be summarized as follows:

Step #1	Dissect the problem into sections of similar and related data.
Step #2	Compare each area to already known natural laws.
Step #3	Use the natural laws, obtain a solution for each section.
Step #4	Combine and integrate the solutions so obtained (from step #3) into one comprehensive solution called the First-Order Solution (Also called the Baseline Solution)
Step #5	Resolve the remainder, which cannot be known immediately, by using the known part (and its solution) to arrive at the second-order solution.
Step #6	Refine the second order solution if the desired degree of accuracy is not reached. The process of refinement has to be repeated "n-times," if needed, to bring about a more accurate solution, referred to as the Final Solution, or the "nth-Order Solution."

"Every big problem was at one time a wee disturbance."

—Unknown

These six steps are summarized in the Figure shown on the next page, from which we can clearly see how any complex problem can be solved systematically and with any desired degree of accuracy.

"Science is nothing but developed perception, interpreted intent, common sense rounded out and minutely articulated."

—George Santayana (1863 - 1952)
Spanish Born American Philosopher and Poet

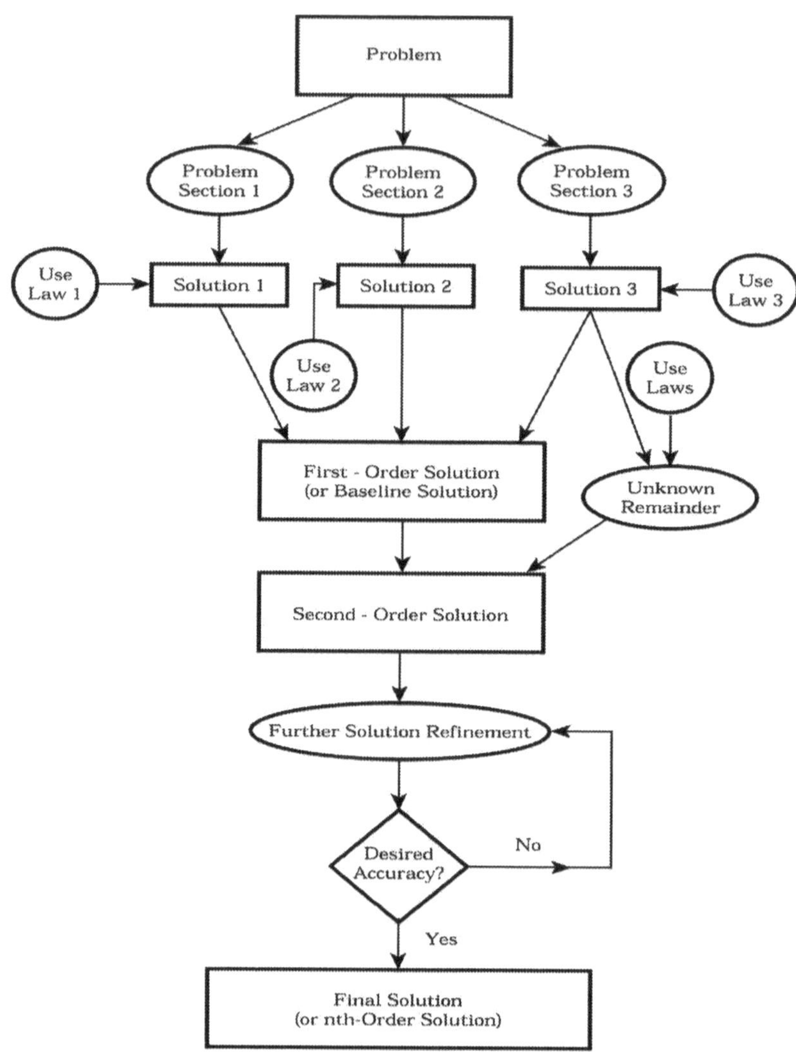

A Flow Chart For A Systematic Method To Solve Any Problem.

The Scientific Methodology

Of all philosophical approaches possible, one has stood the test of time and has emerged as the popular methodology to analyze or to discover new grounds in physical sciences. Its popularity is for obvious reasons since it has put physical sciences on an extremely dominating position with respect to all other "pseudo-sciences," such as political science, social sciences, etc.

The philosophical approach commonly employed by all exact physical sciences is referred to as the "***Scientific Methodology,***" which can be defined as:

"A systematic approach devoted to pure observation as a first step, followed by collection and classification of data (without observer's influence) and formulation of a hypothesis (based on the collected data), testing and experimentation to verify the validity of such a hypothesis and repeating this process for further refinement, until a workable principle is derived ." The scientific methodology is shown in a diagram on the next page.

With this methodology, one can approach any unknown field and gain much workable knowledge about that field and gain a complete mastery of that subject in time. This has been employed over and over in all of our extant and modern physical sciences, such as physics and engineering, with great success.

Such an approach has not been leveled at other fields of study such as art, humanities, politics, economics, etc. Otherwise we would not be experiencing extreme degrees of confusion and turmoil in our financial life such as recession, inflation, and other ills (due to lack of laws in economics) or in constant upheaval and crisis in our society, nationally or internationally (due to lack of natural laws in politics).

"Science may set limits to knowledge, but should not set limits to imagination."

— Bertrand Russell (1872-1970)
English Logician and Philosopher

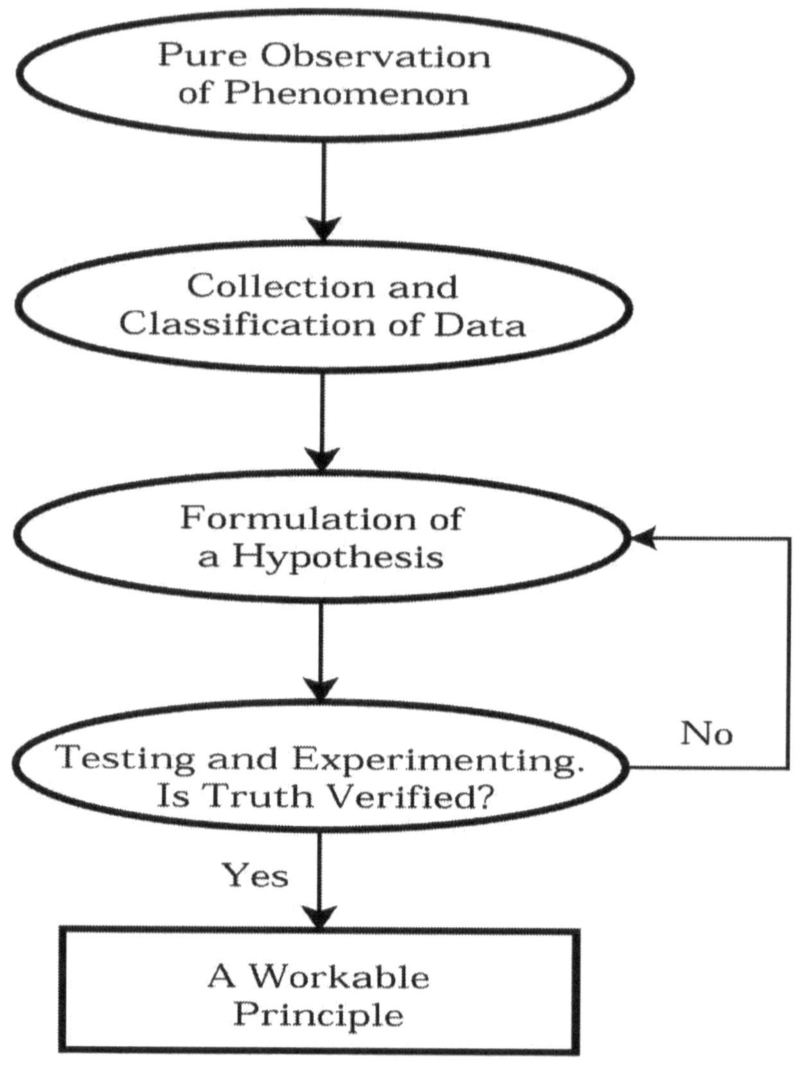

Use Of "Scientific Methodology" To Obtain A Workable Principle.

⊶ ✶ ❁ ✶ ⊷

The Realm of Physical Sciences

Generally speaking, to obtain a physical science, we require first a "Viewpoint" to observe the physical phenomena and then to organize, combine, and integrate the observed phenomena into a unified theory (using the scientific methodology), such that the result is a successful and workable principle under all conditions. The principle that emerges finally should be able to explain all phenomena including past events and observations and should be able to predict future behavior under a set of given conditions as shown below. This is what is required of a physical science!

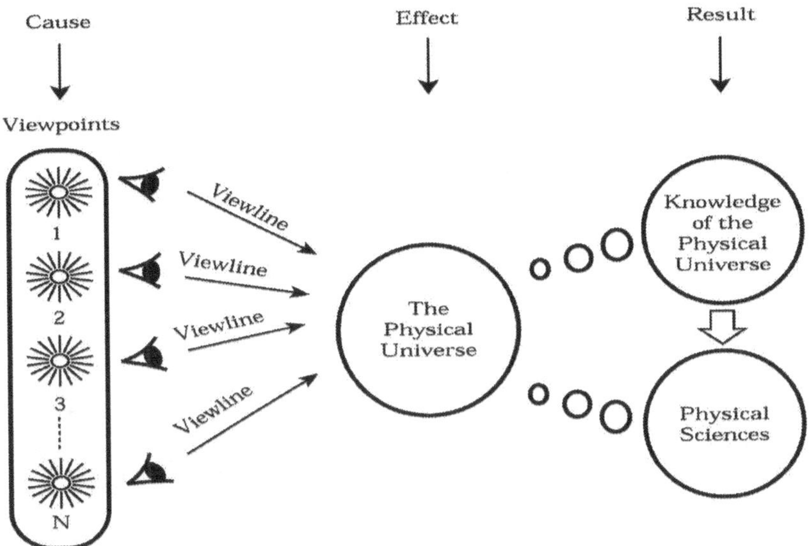

The Relationship Of A Physical Science To The Physical Universe And A Viewpoint.

Let's clearly define the legitimate province of physical sciences particularly physics, engineering, chemistry, etc., by defining them as, "***The study and analysis of the inanimate components of the physical universe, excluding the life force, with the intent of using the principles and results thus found for the betterment (or otherwise) of the conditions of mankind or all life organisms that are symbionts to existence.***" By symbiont, we mean an organism intimately living with another, especially when in a state of mutual advantage (e.g. a pet, etc.).

"Science without religion is lame, religion without science is blind."

—Albert Einstein (1879-1955)
German Born American Physicist, Nobel Prize in 1921

A general living organism can be roughly divided into two components: a) The life force component, which contains within it one or many viewpoints, and b) the body or the physical component. This is a very rough division. Even though these two components are closely intertwined, they are actually distinctly different, each with its own set of laws.

The Figure below shows the relationship of a viewpoint to a physical science relative to animate and inanimate entities.

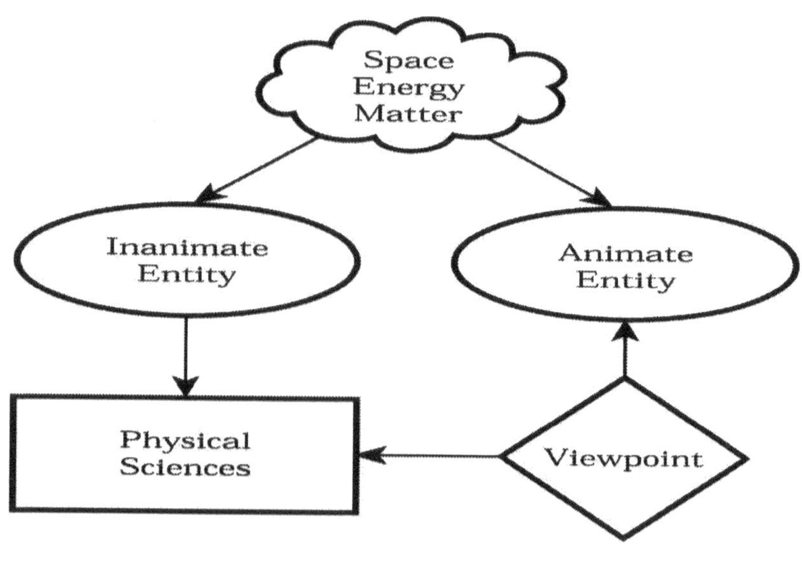

A Physical Science Relative To A Viewpoint.

All of the physical sciences particularly physics and engineering, do not encroach upon the life force component or the biological aspects. They only study the physical component purely, as a mass composed of atoms and molecules.

Clinically, it has been shown and proven that the laws of conservation of energy do not apply to the life force or the thought component of life; thus, the science of physics (which is built squarely on the principle of conservation of energy) fails to predict the behavior of a biological organism.

With this definition in mind, we can see that, for example the field of "waves" focuses on behavior of energy in space, which is but a subset of a much bigger field of study called physics, which is a subset of the physical universe. The latter itself is a subset of a larger sphere of existence called "life and livingness." This means that in the universe of waves, we are not studying an isolated subject but a very narrow field of study which is a subset of another subset (physics), itself being a subset of another larger subset (the physical universe), so on and so forth.

Of course, there are commonality and shared concepts between principles of high frequency electronics and laws of the physical universe. For example, the commonly known principle of inertia (i.e., a moving object tends to remain in motion or a stationary object tends to stay motionless unless acted upon by a force) may be seen to have applications in the current flow in an inductor, wherein the current tends to flow continuously and resists sudden shifts in magnitude or changes in its direction of flow. Therefore, if we suddenly try to stop the current flow by unplugging the wire of a running equipment, we cause sparks at the electrical outlet, which is an indication that the current resists being stopped and actually "jumps" through the air to make a connection to the wire in the wall.

"Science can only ascertain what is, but not what should be, and outside of its domain value judgments of all kinds remain necessary."

—Albert Einstein (1879-1955)
German Born American Physicist, Nobel Prize in 1921

⊶ ☙ ✶ ❁ ✶ ❧ ⊶

The Role of Philosophy in Modern Sciences

It should be noted that any existing field of study is heavily influenced by the philosophical approach and the viewpoint of its founders and pioneers.

The postulates, assumptions, methods, and techniques that they utilized to acquire the foundation material of that science is extremely important in the workability of that science.

"The religion that is afraid of science dishonors God and commits suicide."

—Ralph Waldo Emerson (1803-1882)
American Essayist

Ordinarily, the philosophy of the subject precedes many years, sometimes even thousands of years before its final workable form. For example, the field of digital systems is based upon the two-valued logic system, which was proposed over two thousand years ago by its Greek originator Aristotle and has now come to full maturity with the advent of very sophisticated computers and digital equipment, which accomplish very complex tasks.

"True philosophy invents nothing; it merely establishes and describes what is."

—Victor Cousin (1792 - 1867)
French Educational Reformer, Philosopher and Historian

The Success of Physical Sciences

The very recent emergence of physical sciences over the last four hundred years, particularly electromagnetism as the mother science of many other fields of study, has put the social sciences and humanities particularly subjects such as psychology, sociology and psycho-analysis, etc. in a precarious position. The latter have not been able to provide rigorous and exact results that the former have been able to produce.

Many riddles and mysteries of our lives have been resolved and the whole face of our civilization has been given a major uplift. New tools, machines, and technology have made earth mankind's very own backyard, with new scientific applications and explorations cropping up everyday.

"What is a scientist after all? It is a curious man looking through a keyhole, the keyhole of nature, trying to know what's going on."

—Jacques Cousteau (1910-1997)
French Explorer

The reason for all this excitement and renewed interest in the physical universe lies in the method of approach in analysis and design. By that we mean the systematic methodology the scientists have utilized in the analysis of different aspects of the physical universe. This systematic methodology can be summarized as:

I. First, the founders identified the **four primary postulates** that have gone into building any piece of the physical universe. These postulates are axiomatically true regardless of the nature of the problem. These are a) The postulate of space, b) The postulate of energy, c) The condensed version of (b), which is the postulate of matter, and finally d) The postulate of time, which is the time-track of the interactions of mass and energy in space.

II. Having formulated the four postulates, a series of **axioms** that are the common denominator of these fundamental postulates have been obtained. These developed axioms convey the relationship between the primary postulates and thus express a higher level of understanding about the physical universe at large.

III. Next, a series of **natural laws** were extracted that would precisely express the relationships between the forces and energies involved as a function of time both qualitatively and quantitatively.

IV. Then specific and exact **units** were adopted to facilitate the process of quantifying the four basic entities (such as meter, Joule, Kilogram, and second), which were followed by development of advanced measuring techniques for all quantities associated or derived from these four basic postulates and entities such as power (Watt), current (Ampere), etc.

V. Furthermore, an exact set of **nomenclature** describing precise phenomena has been developed. Using this developed set of nomenclature and specific terminology, the understanding of the discovered principles and their communication to others were increasingly facilitated and moreover, the road for future discoveries was superbly paved.

VI. Next, with the help of an abstract language such as **mathematics** a series of exact theorems have been shown to exist. The set of mathematical theorems propose identities in results and simplify a great many complex problems, abstract or physical.

VII. Following these developments a series of **tables and charts** have been developed for rapid analysis and design of problems related to the physical universe.

VIII. Use of computer and the developed application **software** programs to simulate and solve idealized problems, have added a whole new dimension to the accuracy and speed of obtaining solutions under idealized conditions.

IX. Finally, with this much understanding about the physical universe, an onslaught of new **machines and systems** has accompanied this mental triumph

Steps I-IX are diagramed below, where the postulates and axioms are at the heart or center of the circle of operation, and other activities systematically progress outward and toward the most outer layer of the circle, the application mass, which is the most visible part of any science.

"Science has proof without any certainty. Creationists have certainty without any proof."

—Ashley Montague (1905-1999)
English Anthropologist

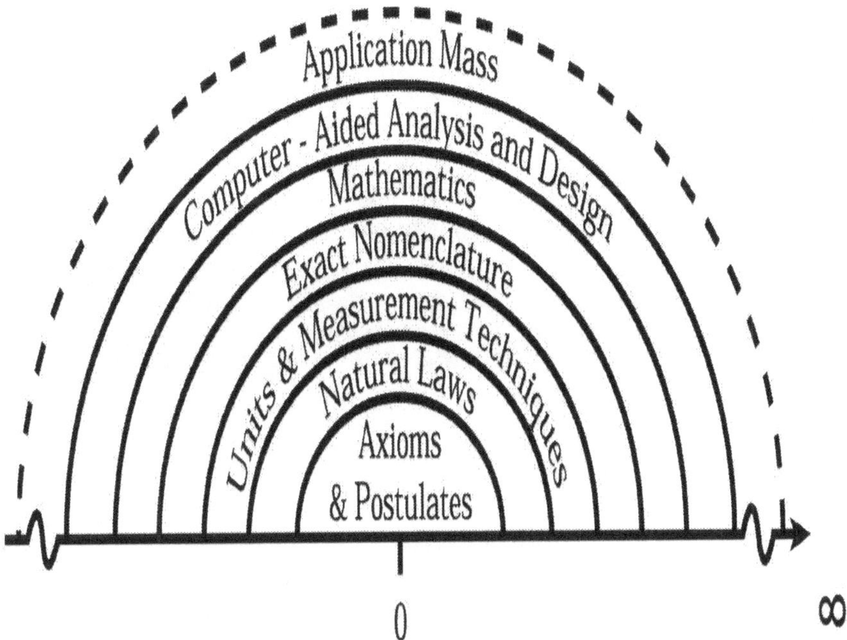

Progression Of A Science From Postulates To Application Mass.

⊷ ✶ ❁ ✶ ⊶

The Social Sciences

"Just because it's common sense, doesn't mean it's common practice."

—Will Rogers (1879-1935)
American Actor-Entertainer, Philosopher and Journalist

The methodology utilized in the development of physical sciences is quite different than the one used in the development of social sciences. Thus the same cannot be said about "soft sciences" such as "social sciences, political science, psychiatry, and humanities as a whole." One should be alert that even though the word science is used in their titles, but they are actually "pseudo sciences," for reasons explained below.

"I believe that a scientist looking at nonscientific problems is just as dumb as the next guy."

— Richard Feynman (1918-1988)
American Theoretical Physicist

First, there have been no discovered postulates about life force, its appurtenances or symbionts. As a result no natural laws or axioms could have possibly been discovered or derived that can generally and uniformly be applied to all life forms.

Moreover, the nomenclature developed in these social sciences is very elementary and does not adequately cover fundamental or advanced phenomena.

Because of the lack of exact fundamental laws or nomenclature, workers in the field have not developed any axiomatic approach to achieving unique solutions to problems, nor have they developed any workable charts and tables categorizing states of beingness of an individual or a society in order to diagnose and solve an individual's personal problems in life or social ills of a society or potential problems facing man, in general.

They have stagnated in a deep quagmire of confusion with no end in sight. The Figure below shows the development of the social sciences and their degree of workability.

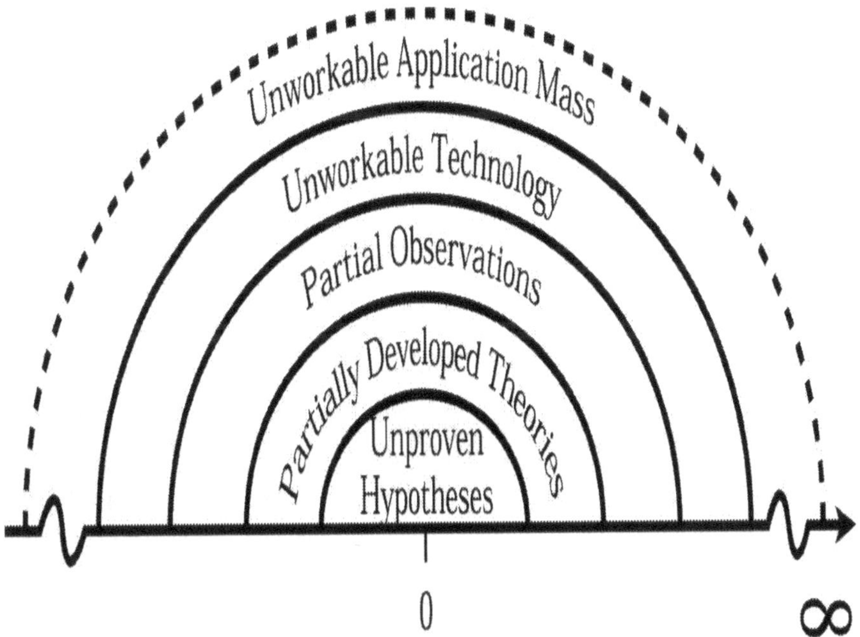

Progression Of Social Sciences From Unproven Hypotheses To Unworkable Application Mass.

Their use of computers and software is primarily for bookkeeping and statistical purposes and does not aid in providing rapid and accurate answers that life poses to life organisms, as individuals or groups.

"Knowing others is wisdom, knowing yourself is enlightenment."

— Lao Tzu (600 -531 BC)
A Chinese Philosopher

A vast number of modern theories about the nature of human beings have been proposed in psychology and sociology and other related social sciences over the last hundred years, from the days of Freud starting with his famous theory of psychoanalysis to now (1895-present) and has created a rather shifty subject, with many fads abounding on a year-round basis.

These theories are founded upon premises and postulates that cannot stand the acid test of "*Scientific Methodology*" earlier discussed. They do not stem from fundamental truths, thus they change every few years. No new discoveries in the subject that could have led to some major breakthroughs in the subject have been uncovered or reported.

In short, these subjects are not true sciences in the exact sense of the word but can be classified as "pseudo sciences" at this point of time. They have been stagnant over a hundred years and can be comfortably referred to as theoretical approaches at best.

"The fool doth think he is wise, but the wise man knows himself to be a fool."

—William Shakespeare (1564 - 1616)
English Dramatist, Playwright and Poet

Half-Truths in Sciences

It can be observed that much false data can enter into a subject by unscrupulous workers in the field based on their opinions or misunderstandings of the subject. As an example, let us consider a physics professor who, while instructing the students on light phenomena, tells them that "a photon has no mass." Such a statement as innocent as it may be, introduces a false datum into the student's mind. We know well from the discussion of relativity of knowledge that absolutes (zero or infinity) cannot be reached in the physical universe. Therefore a zero (0) mass for an actual physical entity is impossible!

What the professor should have said instead should have been something on the order of: "mass of a photon is negligible comparable to other similar particles such as an electron (m_e=9.1095x10^{-31} kg) or a proton (m_p=1.67264x10^{-27} kg); thus for all practical purposes it may be assumed to be zero." Knowing the full data, we can then form our own opinion and realize that a zero value for the mass of a photon is a useful assumption with a certain degree of workability within a limited range.

"The most dangerous untruths are truths moderately distorted."

—George Christopher Lichtenberg (1742 - 1799)
German Physics Professor and Scientist

For example, the mass of a photon, which depends on its wavelength, can be calculated for an infra-red photon traveling in a fiber optics cable having a wavelength (λ=1.55 μm=1.55x10^{-6} meters), to be:

m=1.42x10^{-36} Kg

Thus we see that the mass of a photon at a wavelength of 1.55 μm is approximately 1.56x10^{-6} times (or about one and a half million times smaller than) that of an electron (m_e=9.1095x10^{-31} kg), but it is not zero!

There are many other examples one can think of where this same situation occurs almost on a constant basis. This is quite prevalent, particularly in social sciences where there are no workable series of postulates, axioms, or natural laws, and their textbooks are fraught with unproven theories, authoritative opinions, and arbitrary factors and data. They have no concourse or familiarity with the scientific methodology and thus have not developed a technique by which one obtains unique solutions to any problem.

Moreover, there are a large number of unproven and arbitrary theories, which form the foundation of humanities and social sciences. This is because they are not using the scientific methodology to obtain workable data. Not only that, they also lack a solid scientific series of postulates. Thus, the subject appears chimerical and shifts dramatically every few years.

"Wisdom is what's left after we've run out of personal opinions."

—Cullen Hightower (1803-1878)
Irish Clergyman

Understanding Physics

The viewpoint of physics, accumulated through thousands of years, has shifted considerably through time. It currently has set itself up to assume the viewpoint that views the space of the whole physical universe and has the task of understanding it from the atomic level to planetary, to galaxy level and beyond. It has several divisions, such as ***Elementary Physics, Classical Physics, Quantum Physics, Particle Physics, Astro-Physics***, etc.

From the viewpoint of classical physics' practitioners, everything is certain and can be precisely analyzed or determined with a very high degree of accuracy.

However, from the viewpoint of the quantum physicist, everything is uncertain, and only a probabilistic analysis would show the probability of the occurrence or existence of any event. This is because the particles and energies within the created space only exist on a probability level; therefore, all of the events or actions concerning these can only be expressed as a probability.

Obviously, these two viewpoints of the classical and quantum physics are opposed to each other. These diametrically-opposite viewpoints have come about because the space in which one places the particles and energies is well defined and exact —both physically and mathematically (in classical physics), whereas on a quantum physics level, the space is defined in terms of a number of particles that only exist on a probabilistic level. Furthermore, the mathematics involved is quite indeterministic and all of the obtained solutions are based on a "probability of existence or occupancy" type of proposition.

The only way that these two clashing views of the same physical universe can be compromised is through the observation that fundamentally and microscopically there is an inherent uncertainty in the physical universe. This is caused by the confusion and clash of energies.

However, when a large number of these uncertainties are grouped together to form an object or when life force imbues matter to form a biological organism, then on a grouped and average basis, we have a much higher degree of certainty of analysis (classical physics level of high accuracy) compared to the analysis of a single microscopic particle (quantum physics level of uncertainty).

The collection of many uncertainties of many particles (such as electrons and protons) into atoms and molecules to form and eventually create an object is an apparent certainty that classical physics triumphantly presents in its elegant mathematics. This is a certainty obtained and based on an apparency of many "particles' uncertainty."

By apparency we mean a phenomenon or a condition that appears or seems to be a certain way, which is not the same and really different from its actuality. For example a spoon in a glass of water appears to be broken at the interface. This is an apparency, which differs from the actuality of the situation. The actuality is that rays of light undergo refraction at the air-water interface and slow down as they enter the water. This causes the rays to bend at the interface and thus make the spoon appear to be broken.

Therefore, physics then is based upon a composite viewpoint, where in order to study each of the two subjects one needs to know the proper viewpoint of the originator of the subject and then study the postulates made by the viewpoint. For example the postulates of the quantum physics are as follows:

1) Existence of a microscopically small "created space," which is uncertain.
2) Existence of fast-moving particles, which are difficult to trace but are associated with a probability of existence or a probability wave packet.
3) Existence of the Heisenberg uncertainty principle.
4) Conservation of energy-mass system as a whole where mass and energy are inseparable in terms of nature of origin but convertible.

5) All calculations and results are to be expressed by a probability value and never with 100% certainty.

To understand the uncertainty and complexity that this series of postulates has introduced, a large class of sophisticated mathematical models and equations have been developed to bring about some level of certainty and comprehension to this uncertainty.

Complexity at this level of mathematics is totally unsurpassed in all of physics, and the effort here is trying to bring about a comprehension of a total confusion on a symbolic level.

In this type of confusion, the particles are not only un-locatable but also have indeterminate velocities. Furthermore, they have uncertain amounts of energy, depending on the length of time of observation.

Such a state of affairs has been gilded and made palatable by statistical analysis and assignment of probabilities, which at once makes the scientist comfortable because it is no longer a qualitative but rather a quantitative measure of the "probabilities of existence" involved.

So, we can still think through this confusion by expressing it in terms of equations that are as complex as the quantum mechanical particles themselves.

It is interesting to note that perhaps one of the problems associated with quantum mechanics is lack of a stable and unshakable datum that is totally certain. Such a stable datum could have been achieved, for example, by defining and postulating a space that is clearly delineated and defined. By defining such a space, one would be able to place particles and energy waves and thus would be able to shed off a lot of added uncertainties about the behavior of particles and waves associated with quantum physics.

"The observer, when he seems to himself to be observing a stone, is really, if physics is to be believed, observing the effects of the stone upon himself."

— Bertrand Russell (1872-1970)
English Logician and Philosopher

Mathematics

It is an interesting commentary on the subject of mathematics that needs to be addressed at this stage of our work. Mathematics, at this stage of development, is so interwoven with sciences that it has almost become an inseparable partnership between the two, or so it seems.

Upon further investigation, we can make the following observation: *Mathematics are short-hand methods of stating, analyzing, or resolving real or abstract problems and expressing their solutions by symbolizing data, decisions, conclusions, and assumptions* (see the Figure shown below).

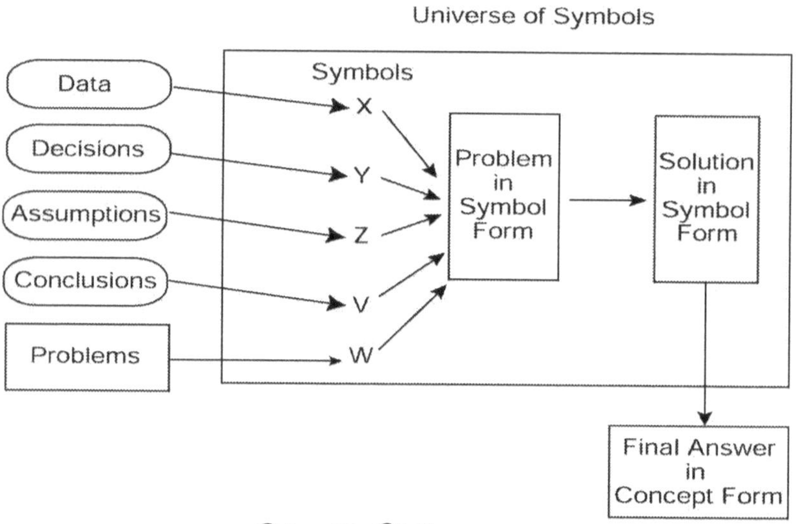

Use Of Mathematics.

This means that a science exists as an organized body of data expressed as a series of laws, concepts, etc. with mathematics only as a servo-mechanism to expedite its organization, conveyance, and communication through symbolic representation. In other words, mathematics is junior in importance to the science it serves.

"Except in mathematics, the shortest distance between point A and point B is seldom a straight line."

—Anonymous

As a **Point of caution**, *we should note that utilizing exact mathematics in expressing natural laws or solving problems related to the physical universe introduces an absoluteness into the final solution which contradicts rule of non-absoluteness of the physical universe data. This contradiction can be resolved by realizing that one should only expect an approximate correlation between the actual measured quantities and the final solutions obtained through the use of exact mathematics.*

As an example, let's see how we would express the income of an engineer, mathematically?

To answer, we will consider a first order of approximation, where one may considers his income (I) to be proportional to and depend upon:

a) His work ethics and honesty (H),
b) Correct actions (A),
c) Hard work (W), and
d) Knowledge of the subject (K)

Thus, one can write the following relation as a shorthand:
I=H.A.W.K

So, this simple equation will serve as a short-hand for all the above information. It shows a linear relationship between one's income and the above parameters on a first-order of approximation, and can be used as an abbreviated version of the *above datum, but never to replace the datum itself.*

" The strongest arguments prove nothing so long as the conclusions are not verified by experience. Experimental science is the queen of sciences and the goal of all speculation."

—Roger Bacon (1220-1292)
English Philosopher and Scientist

⊶ ✶ ❁ ✶ ⊷

Sciences as a Subset of Life

Sciences are organizing tools that, through a systematic process of observation of the chaos of the physical universe, try to uncover the governing principles, and in doing so, intend to bring about comprehension on a mental plane and a higher level of order on a physical level. Such a higher order on a physical plane is accomplished generally through the invention of intricate devices or systems, or the creation of new application mass.

"To read means to borrow; to create out of one's readings is paying off one's debts."

—George Christopher Lichtenberg (1742 - 1799)
German Physics Professor and Scientist

Since life and livingness, as imbued by the life force, is set up in a hierarchical fashion at the top of the pyramid of knowledge affecting the entire pyramid, thus we realize that sciences as a pool of knowledge and a source of accurate information about our world, created by and for the viewpoint, are merely a subset of life as shown on the next page.

From this Figure, we can see that sciences are below the level of the physical universe, which itself is an application mass and thus a byproduct of the exact series of postulates put forth by the viewpoint and the life force at a distant past.

"Science is simply common sense at its best that is, rigidly accurate in observation, and merciless to fallacy in logic."

— Thomas Henry Huxley (1825-95)
English Biologist

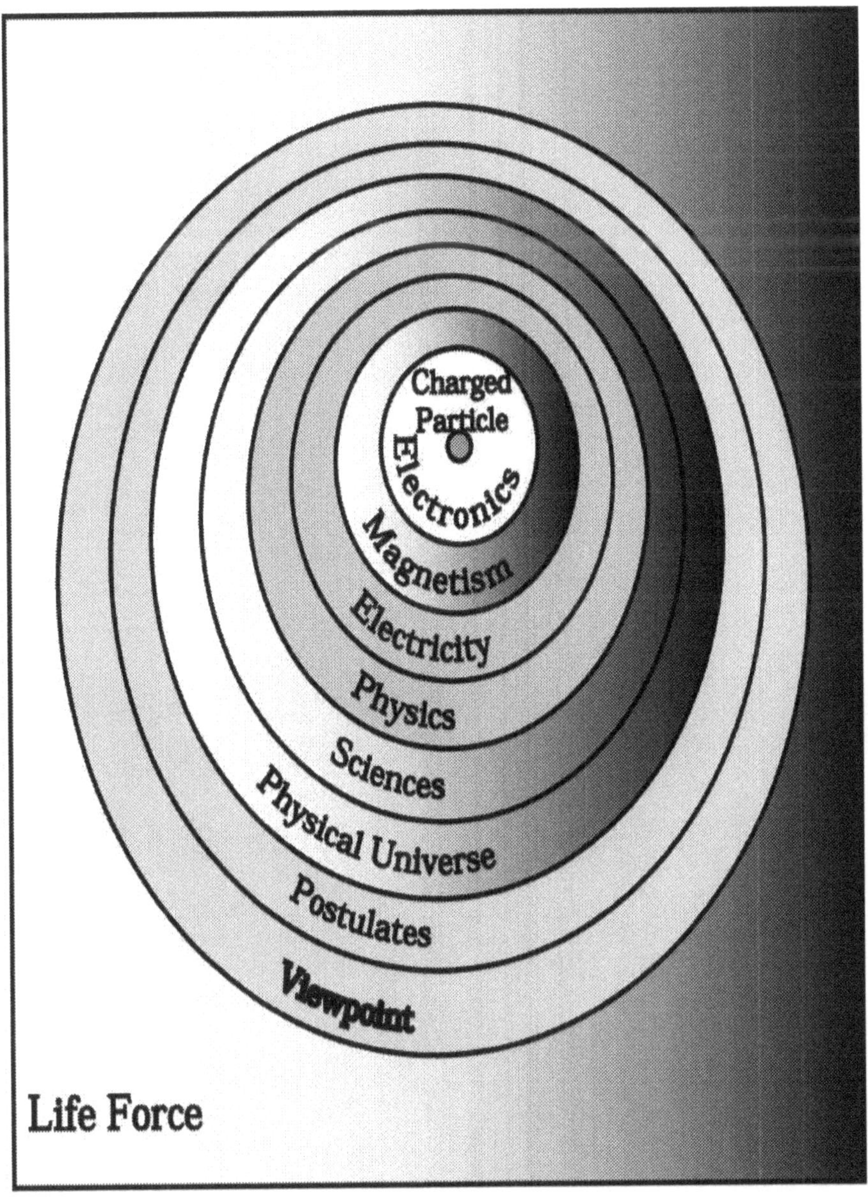

A Diagram Showing Sciences As A Subset Of Life With The Viepoint As Thr Main Orientation Point..

⊶ ꕥ ✶ ❁ ✶ ꕥ ⊶

PART III

The Code Disintegrator

A Missing Vital Element

"Absence of evidence is not evidence of absence."

—Dr. Carl Edward Sagan (1934-1996)
American Astronomer, Searching for Intelligent Life in the Cosmos

There is an implicit consideration that is missing in all of the powerful postulates, and that of course is the essential entity that envisioned and considered the three original postulates, which we commonly refer to as "the viewpoint!"

Through close examination, we can see that to create any universe, and physical universe is no exception, we must first and foremost have a viewpoint. The scientific world has taken this essential element for granted.

"There are things I can't force. I must adjust. There are times when the greatest change needed is a change of my viewpoint."

—Denis Diderot (1713-1784)
French Philosopher

By definition, to create the physical universe, we must have a viewpoint. This viewpoint can be a single viewpoint (or a collection of viewpoints), which postulated some very powerful postulates and thus placed the whole physical universe here and in motion.

Once these primary postulates were made, in order for another viewpoint (such as viewpoint A) to view the physical universe, "viewpoint A" must agree totally with all of the four postulates. "Viewpoint A" must set up one's space coincidental with the created space of the physical universe and then and only then, one can view the energy particles, matter particles and their interactions. At that exact moment, "viewpoint A" is said to be placed in the physical universe!

We are not at this point concerned about the identity of the viewpoint. We are not interested to know the identity of the viewpoint (or viewpoints) or the postulator (or postulators) that postulated the three original postulates, which led to the four primary postulates and neither do we want to know "how come" or "why" this set of postulates. That is not the question before us. What we want to know in this work is: a) What was the exact set of postulates that led to the final construction of the physical universe, and b) How can we use the discovered information to our advantage by improving our life conditions through a better analysis or a more complete design of devices, components, or in general, any engineering system.

"If there is any one secret of success, it lies in the ability to get the other person's point of view and see things from that person's angle as well as from your own."

Henry Ford (1863-1947)
American Industrialist and Pioneer of the Assembly-Line Production Method

The Viewpoint

One of the main abilities of a viewpoint is the ability to assume beingness. In the sciences, it can be observed that there are two grades of beingness that a viewpoint is allowed to assume. With regard to the physical sciences, it is an important point to understand because it leads to resolution of many tangled mysteries hitherto unresolved.

"Our death is not an end if we can live on in our children and the younger generation. For they are us, our bodies are only wilted leaves on the tree of life."

—Albert Einstein (1879-1955)
German Born American Physicist, Nobel Prize in 1921

The classification of the basic categories of viewpoints in the sciences is defined as follows:

A POSTULATING VIEWPOINT: *Is a viewpoint that can postulate something and bring it into existence. In the scientific arena, it ranks the highest and is at the top of the pyramid of knowledge presented in an earlier part "The Essence of Physical Sciences." Such a postulating viewpoint creates space, energy, matter and time and furthermore through observation can generate a series of facts and data and thus a body of knowledge can be born. A simple example on a small scale could be an inventor of a novel device. On a larger scale, an example of a postulating viewpoint would be the founder of a new science (e.g. Faraday, Maxwell, etc.), where a whole system of application mass is postulated into existence [Ref. 8, pp. 1, 31].*

AN OBSERVING VIEWPOINT: *This is a viewpoint dedicated to observation of effects. Such a viewpoint is at the bottom of the pyramid of knowledge, where only the created spaces, created energies, created masses and created times are observed. There is no created knowledge, only observation of phenomena and perhaps* *strict recordation of that observation with no added formulation or organization of the data. A simple example would be an observer reading a meter. On a broader scale, an observing viewpoint would be a scientist reporting on a major discovery as a result of observation of some natural phenomena.*

A student of a science initially falls into the category of "observing viewpoint," since one is observing what has been discovered, postulated and created as application mass. However, eventually the student should be able to apply the obtained knowledge and lift oneself by the bootstrap, up the line in the direction of "postulating viewpoint" and eventually assume or duplicate its viewpoint.

"One who is too insistent on his own views, finds few to agree with him."

—Lao Tzu (600 -531 BC)
A Chinese Philosopher

These two major subdivisions of viewpoints are shown below.

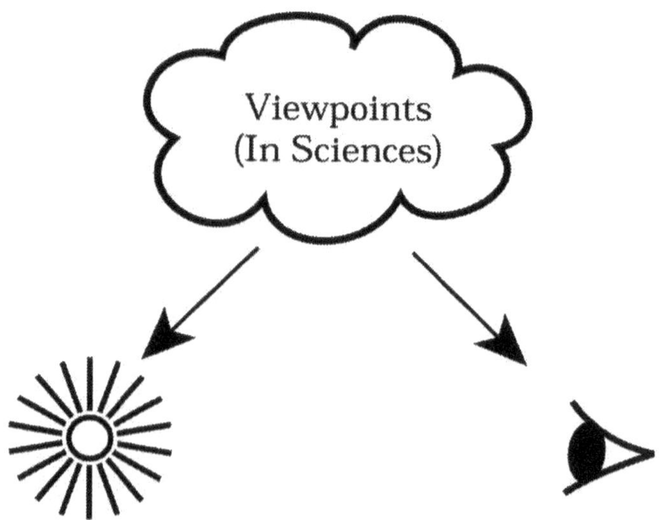

The Two Subdivisions Of Viewpoint.

It should be noted that in actuality there exists a gradient of viewpoints between these two extreme cases. However, these two viewpoints have very definite applications and ramifications in all physical sciences.

To assume such beingnesses as described by the above two categories (of viewpoints) require the viewpoint, in each case, to remain at all times in present time. Digression from present time and delving into the past leads to other complicated beingnesses that are beyond the scope of this work. One example of such a case could be referred to as a "reactive viewpoint," where the viewpoint observes and associates the past experiences to such an extent that any action at present time, which has some past connotation or resemblance is reacted upon.

This action actually adds extra complexity into the present time action of postulating or observing process, as delineated above. In other words, this action will add a series of "arbitrary factors," which will worsen the final outcome of postulation or observation process.

The action of the reactive viewpoint is far different than an analytical and logical approach of using the past results to improve current hypothesis or observations, in that the reactive viewpoint associates irrelevant and illogical past data and facts with the present time data. Such an action will adulterate the data in the direction of worsening the present time problems and their associated solutions.

In this work, wherever we use the word "viewpoint" we mean the first concept, which is a "postulating viewpoint." On the other hand, when we want to use the concept of observing viewpoint, we use the term "observer."

"*Not everything that is faced can be changed. But nothing can be changed until it is faced.*"

—James Arthur Baldwin (1924-1987)
American Novelist

The New Science of Viewpoints

The importance of postulates and the role they play in sciences cannot be underestimated. A subject far more important than postulates, considerations, laws and equations in the sciences is "the viewpoint," either as the postulator of new concepts or products the observer of new phenomena. These are for the most part, much neglected topics and yet form an essential basic in terms of understanding and organizing physical sciences and should not be taken lightly.

The current work deals intimately with the concept of viewpoint and proposes a new subject called "the science of viewpoints." It is a totally new look indeed! It is deceptively easy to miss this point as extant sciences have done so. It is an extremely powerful way to look at the universe, particularly the world of physical sciences, utilizing the concept of viewpoint.

We have sciences today, which claim to be "viewpoint-less" and yet upon close examination we see that they are totally "viewpoint-dependent." The next example clarifies this point further.

"If you can't explain it simply, you don't understand it well enough."

—Albert Einstein (1879-1955)
German Born American Physicist, Nobel Prize in 1921

Let us consider an equation stating a natural law (in the differential or integral form) such as one of Newton's laws, and examine whether it has a viewpoint built into it or not.

Upon careful examination of the equation we see that it has a number of scalar or vector quantities such as force, mass, etc., with a number of space variables such as (x, y, z) and time variable (t). This equation concerns a natural law and at first glance appears to be simply stating a fact of how different variables interrelate with each other in time and space.

However, it is interesting to note that an important fact is built into this equation: in order to get any meaningful answer from it by solving for the unknown variables, we need to know where the reference point for space (x, y, z) and time (t) is; that is to say, where is x=0, y=0, z=0 and when is t=0?

Now, having established the reference point in time and space, we are ready to measure observable phenomena in that particular space and time zone. Moreover, we need to select a proper reference frame to solve this equation correctly. To that end, we need to position the observer (i.e., the observing viewpoint) in the coordinate system relative to the reference point, space-wise and time-wise. This immediately brings the concept of "viewpoint" back into the equation, something that was put out of sight by the mere introduction of the symbols.

Therefore, we can see that there are huge assumptions built into scientific equations, which no one has ever cared to look for or tried to bring to the forefront. This is where the sciences have gone into a corner and trapped themselves into a complete stagnancy. The above scenario of apparent "viewpoint-less complacency" and yet "totally-dependent on a viewpoint" in actuality applies to all scientific subjects, equations, and laws.

"We can't solve problems by using the same kind of thinking we used when we created them."

—Albert Einstein (1879-1955)
German Born American Physicist, Nobel Prize in 1921

This could be considered to be an unintentionally-placed scientific pitfall, which has been with us for centuries and many well-intentioned scientists and researchers unknowingly have contributed to it. It has made the subject of physical sciences rather unpalatable for the average man and has given a deceptive look to sciences, which is most unfortunate.

Einstein made a feeble attempt to generalize these equations by proposing his relativity theories, which revealed the viewpoint dependencies of these equations and made it come to light to some extent. Nevertheless, truth be told, all of the laws of physics have this viewpoint dependency built into them with great fervor!

It is interesting to note that these frailties are never discussed and, in fact, are heavily camouflaged in all textbooks. It is not only undiscussed, but is mostly ignored. It has become a neglected subject to a large extent.

The present work is putting the concept of the "postulating viewpoint" (also sometimes referred to as viewpoint or postulator) as well as the "observing viewpoint" (also known as the observer) back into the scientific laws. This is because the laws of physics are derived systematically from postulates, which they in turn, are derivable from the postulating viewpoint. In other words, we can analyze, solve and understand all scientific phenomena with full observance of these two powerful elements in place and come away with fantastic simplicity and enormous understanding.

The new subject is not the same as the old physics but is a totally new approach in dealing with the physical universe. It actually generalizes all of the old physical concepts and adds tremendous simplicity to the material. This new approach to the sciences can be called the new **"Science of Viewpoints."**

In short, the science of viewpoints is a new approach to understanding scientific postulates, concepts, laws and considerations. This approach not only simplifies, but generalizes the subject of physics and engineering thoroughly. It adds meanings to the basics where there were none.

Considering and utilizing the ever-important concept of duality, which has been interwoven throughout our work, we can conclude that a "postulating viewpoint" and an "observing viewpoint" are a pair of dual concepts in the general subject of viewpoints.

"Insist upon yourself. Be original."

—Ralph Waldo Emerson (1803-1882)
American Essayist

The Postulation of a Viewpoint

The existence of a viewpoint is an important assumption. The assumption of a viewpoint unconditionally is actually equivalent to adoption of a new and yet very workable postulate: "Let there be a viewpoint!" This postulate has never been violated on a smaller scale, under any and all conditions, and is constantly and continually being used in engineering sciences to design new devices, machines, etc.

The existence of such a viewpoint on a "grand scale" at this point in time is purely a theoretical assumption, and is adopted here to facilitate our work because it has deep roots in the sciences on a smaller scale. This postulate gives alignment to the physical sciences and makes it a more understandable and palatable field of study.

The postulation of the existence of a viewpoint on a large or grand scale is assumed to be theoretical and is purely for the sake of alignment of sciences at this stage. To provide "actual proof" of the existence of a viewpoint (or a collective viewpoint) on a grand scale is well beyond the scope of this work and is actually not the focus of this book.

"One's first step in wisdom is to question everything - and one's last is to come to terms with everything."

—George Christopher Lichtenberg (1742 - 1799)
German Physics Professor and Scientist

Furthermore, we need to realize that there is no physical science extant today that is prepared to prove or disprove the concept of viewpoint or delve into a discourse on the subject, because this concept *implicitly* and yet uniformly has been adopted as a postulate, and as such, no one has ever questioned it on a smaller scale!

This is an important point to grasp because we do not intend to delve into deep, philosophical discussions; we leave the onerous task of "proof of viewpoint's existence" to knowledgeable philosophers and religious scholars. We do not intend to prove the concept of viewpoint but to postulate its existence (i.e., assume it to be true without proof and use it as a basis for reasoning) and then utilize it to bring a tremendous simplicity and alignment to our ensuing discussions.

Moreover, the reason for the postulation of a viewpoint on a grand scale is not an *arbitrary factor* because we have proof of its actual truth and existence on a small scale, and we know how much understanding and simplicity it brings into analysis and design of engineering problems.

Its postulation on a grand scale, even though theoretical at this stage, could be considered to be purely an advanced "scientific extrapolation" of a principle observed to be unconditionally true at a lower scale, which has remained inviolate through the ages. What has been done here is that we have generalized the "small scale" concept of viewpoint and inferred a higher truth by "inductive logic" and arrived at a powerful postulate. This postulate allows us to bring about a tremendous amount of workability, simplicity, and understanding to the sciences.

The proof or disproof of the existence of the viewpoint is actually immaterial, because if truth be told, we have adopted the existence of the viewpoint as a "postulate," which by definition is "an assumption which requires no proof and is held to be true unconditionally and for all times." This means that it is assumed to be true at all times purely by agreement, and perforce is an absolute. Therefore, the proof or disproof of its existence is totally unnecessary for our work and would not change any of our discovered principles the slightest bit!

"Those who cannot change their minds, cannot change anything."

—George Bernard Shaw (1856-1950)
Irish Writer, 1925 Nobel Prize for Literature

The Universal Viewpoint

In view of the existing status of the sciences on earth today, the subject of viewpoint as set forth in this book brings out a whole new perspective on sciences.

It is relatively easy to see the function of the viewpoint when dealing with engineering design or analysis. In other words, it may be relatively obvious to see that before there was a manufactured machine, there was the concept of the design (i.e. the postulate) of the machine in the designer's mind.

The concept of the design indicates that before there was the space of the machine, there was the postulate of space. Before there was the mass of the machine, there was the postulate of mass. Before, there was the energy particles running in the machine, there was the postulate of energy and before the machine was put to use, there was the postulate of time built into the machine, which determined the useful lifetime of that machine.

However, the role of a viewpoint in the creation of a whole planet, a solar system, a galaxy or in general the whole physical universe may not, at first glance, be so evident. This is where we need to resort to inductive logic, and based on the evidence on a smaller scale, assume or better yet postulate the existence of a viewpoint (or collective viewpoints) on a larger scale, which has caused the creation of what we call the whole physical universe.

It should be noted that the existence of a general viewpoint (that creates space, energy, matter and time) is not a matter of belief or faith in the deities or the Supreme Being. This concept is a postulate, which is simply an extrapolation of a lower scale observation.

"Before God we are all equally wise - and equally foolish."

—Albert Einstein (1879-1955)

German Born American Physicist, Nobel Prize in 1921

⊶ ✶ ❁ ❁ ✶ ⊶

The Origin of Viewpoint

So far, we have examined that a viewpoint can set up a universe by postulating a space within which energy forms and masses in relative motion are placed to cause the creation of a universe.

However, we need to consider a new factor called the "life force," whose relationship to the viewpoint or the physical universe needs to be clarified at this stage.

The English Dictionary defines life force (also called **élan vital**—a French phrase meaning “living force or energy”) as, "***The creative force within a live organism, which is producing the life energy and is capable of building and growing physical structure; it forms the substance of consciousness.***"

With this definition in mind, it can be observed that the life force is capable of setting up a viewpoint, which is innate to the living organism (called the innate viewpoint). The innate viewpoint uniquely represents the living organism or the life form, and is capable of creating space for him.

“Let go of your attachment to being right, and suddenly your mind is more open. You're able to benefit from the unique viewpoints of others, without being crippled by your own judgment.”

— Hedley Ralph Marston (1900-1965)
Australian Biochemist and Physiologist

Moreover, the life force is capable of artificially creating and assuming many viewpoints, which are not innate to the organism (called artificial or assumed viewpoints). An example of this would be an actor, who by assuming a role, takes on the beingness (or viewpoint) of a particular character in a play and portrays that viewpoint in different life situations and settings. The artificial viewpoint is equally capable of creating a space, which is unique to that character and distinctly different from the space created by the innate viewpoint.

The key point to realize here is that the life force feeds all of these viewpoints and thus is the actual energy source powering up any given viewpoint. Special care must be taken here not to confuse the two and assume that they are synonymous. They are not! Life force is far superior to the viewpoint and not vice versa as shown below.

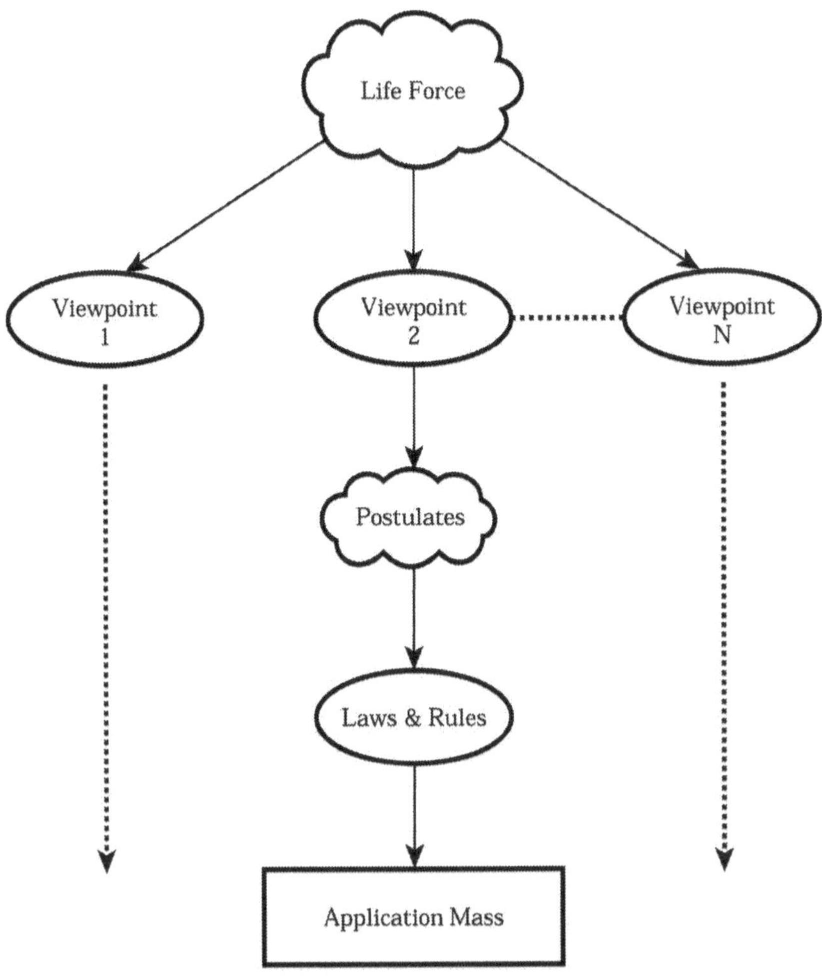

The Relationship Of Life Force To A Viewpoint And Its Byproducts.

"The ability to accept responsibility is the measure of the man."

—Roy L. Smith (circa early 1900s)
American Author

The process of "creation of space" by a viewpoint has a dichotomy in the physical universe. The dichotomy of "creating space" is "created space" in the physical universe, which can be observed in terms of the created space of an object, a device, and so on.

There is a certain degree of parallelism that exists between the parts of the physical universe (such as energy, matter, etc.) and its counterparts created in the mind or body, which are the byproducts of the life force in association with the physical body. The combination of the life force and the physical body, forming a "biological organism," is commonly referred to as a "life unit." An example could be the "energy" that a life form creates in moving an object, which has similarity to the physical universe energy but is distinctly different than its physical counterpart.

"You yourself, as much as anybody in the entire universe, deserve your love and affection."

— Buddha, '*The Enlightened One*' (563-483 BC)
The Title of Indian Prince Gautama Siddhartha, Founder of Buddhism

The concept of parallelism of the life force and the physical universe cannot be extrapolated indefinitely to a point of considering the "life force" and the "physical universe" to be the dual of each other. This concept at first glance may appear to be correct, however, it is an erroneous conclusion since the two are not of comparable magnitude! This is so, because the physical universe (such as an application mass of a device) is far junior to the life force, which created it. That is to say, any part of the physical universe can be derived from a postulating viewpoint, thus, the two are not dual. In short, one is the "creator" (or total cause) and the other is the "created" (or total effect).

"More gold has been mined from the thoughts of men than has been taken from the earth."

—Napoleon Hill (1883-1970)
American Author

The Qualities of Viewpoint

"Those who are free of resentful thoughts surely find peace."

— Buddha, '*The Enlightened One*' (563-483 BC)
The Title of Indian Prince Gautama Siddhartha, Founder of Buddhism

Having agreed that the concept of viewpoint is purely a postulate (or a theoretical assumption held to be unconditionally true) at this juncture, we assign to it certain specific qualities, abilities, and properties.

We may assign numerous qualities to a postulating viewpoint on a grand scale (or small scale), amongst which some are more applicable to our work than others. Recalling the earlier discussion on "How to Create Any Universe," we can conclude that a postulating viewpoint has the ability to:

1. **Create space,**
2. **Create energy,**
3. **Create matter, and**
4. **Create objective and subjective time**

With this theoretical assertion, which is actually a "scientific extrapolation," we have now aligned and connected the field of physical sciences (dealing with inanimate matter) to a much larger sphere of existence called "life," which deals with many things including inanimate or animated matter and above that the power of postulation as shown on the next page.

Special care must be given here to the word "create," which in our work means, "To bring into existence, to make or design something requiring skill, innovation, etc." The definition of the word "create" as intended here excludes the concept of creation by higher powers such as God or supernatural entities.

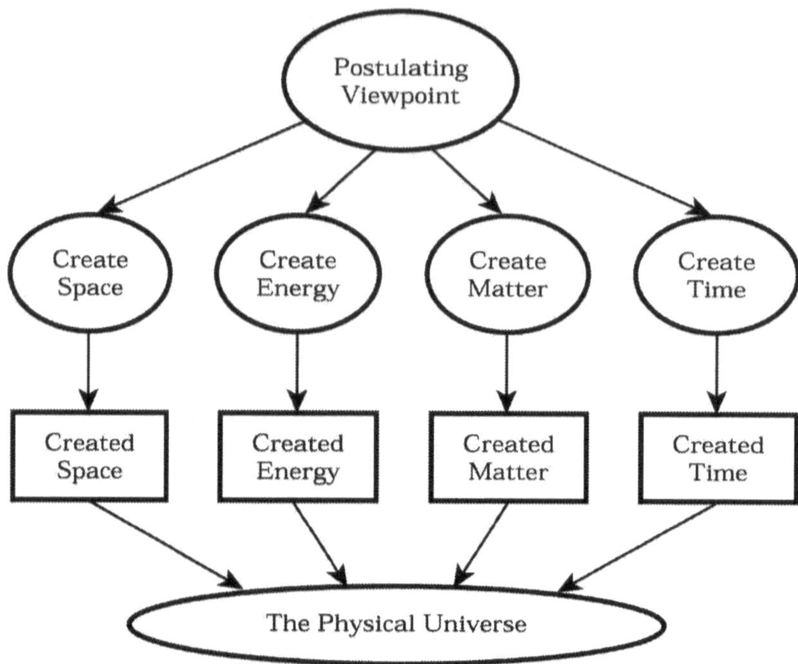

The Process By Which A Postulating Viewpoint Creates The Physical Universe.

There is a **caveat** we need to consider here and now. In proposing the concept of existence of a viewpoint necessary for the creation of any application mass, we have entered the forbidden ground, the land of never-never, where no one has been allowed to walk since mid 1500-A.D. This is where sciences started to go about discovering the principles of the physical universe with total denial of a viewpoint in their literature, and thus created a viewpoint-less subject, which is a total apparency.

They discovered monumental things in this manner, but in this process totally forgot where Man was traveling on a mental plane. This neglect surely brought forth many machines and gadgets and thus a more sophisticated world to live in, but it carried a hefty price tag of leaving behind the majority of the public in a dust of confusion.

The average man cannot think easily with the scientific materials. The heavily symbolized language of mathematics is not easily palatable to the average intellect and thus all of the great discoveries that have accumulated through time are locked up in a symbolic language, with which the majority have no concourse or familiarity. This is the state of sciences that we find today. To change this undesirable condition and make sciences connect with the rest of the humanity and the world at large, which is heavily viewpoint dependent, we have to assume a shunned word in sciences and that is a "viewpoint."

The introduction of the concept of viewpoint (even if on a theoretical level) is necessary if we are going to salvage the stagnant sciences of the physical universe, which do not recognize the postulate as the "cause point or source" and actually stand in awe of a " total effect" called the application mass.

"Whatever you cannot understand, you cannot possess."

—Johann W. von Goethe (1749-1832)
German Poet and Novelist

Mankind's Genesis

It is interesting to note that physical presence of mankind on this planet, purely as a biological form, dates back to several million years ago. In fact, there have been many species of Man on earth since about three and a half million years ago but all are extinct now except for one species.

In other words, none have survived to the current times nor have they had the success or the intelligence to tame the savage forces of nature except for the modern man, a species scientifically referred to as the Homo sapiens (literally, the knowing man). Therefore, when we say "modern man" or "human being," we specifically mean the only living species, "Homo sapiens."

Homo sapiens appeared on earth approximately 35,000 to 100,000 years ago and as a species has made steady progress toward a higher level of existence in terms of survival potential, especially over the last 450 years.

Therefore, we can see that, while the previous species of Man were defeated by the ferociousness of the physical universe due to their lack of ability in handling the problems that the physical universe handed them, the present species of Man (i.e. Homo Sapiens), is being defeated by another set of factors, namely, his own ignorance about himself and a lack of genuine knowledge of how to handle others on a societal and global level. This factor alone puts Man on the endangered species list, since the current state of science and technology on earth clearly indicates that the physical universe has already been defeated many times over!

"We may have different religions, different languages, different colored skins, but we all belong to one human race."

—Kofi Annan (1938-)

Ghanaian UN Secretary-General, 2001 Nobel Peace Prize

The Endangered Species

We can undeniably add Man to the endangered species list. Specifically, the reasons for adding man to the black list of creatures practically on the verge of extinction and moving toward the state of nonexistence, are rather obvious and simple if one cares to look them over.

First and foremost, there is a lack of a genuine administrative and group science and technology dealing with the handling of life on a society level as a whole providing intelligent solutions. Instead there is the handling and solving of the sociological or international problems through the use of destructive forces (i.e. use of wars, utilization of weapons of mass destruction, extermination of one's own kind, etc.).

Secondly, there is a lack of a true science concerning life along with its attendant technology for providing intelligent solutions to problems on an individual or a personal level, where an average member of the Homo sapiens finds himself fighting the forces of the element and other life forms on a constant struggle for survival. Instead we have solutions which are, again, all brute-force dependent.

These solutions can be classified under the headings of a) Chemical solutions (i.e. poisons, drugs, etc.), b) Use of electric energy to bring forth shocks, radiation, etc., to alter one's physical or mental state, and c) Use of forceful methods of elimination and removal of any undesirable areas or conditions of the individual's physical body (i.e., surgery, cautery or burning, etc.).

"A man's character may be learned from the adjectives, which he habitually uses in conversation."

—Mark Twain (1835-1910)
American Humorist and Writer

All of these solutions, whether on a personal, society, or international level, are force-based and have mostly proven to be unworkable since they interject a new arbitrary factor (i.e., unintelligent force) and thus worsen the whole problem, as will be discussed shortly.

"Education is a human right with immense power to transform. On its foundation rest the cornerstones of freedom, democracy and sustainable human development."

—Kofi Annan (1938-)
Ghanaian UN Secretary-General, 2001 Nobel Peace Prize

Of course, there are other incorrect handling or solution areas that one may consider, but the above categories cover the most significant problematic areas, which are pushing the species of Homo Sapiens to the brink of extinction at a rapid pace, just like all other species of man, which have vanished forever because they were facing problems for which they had no "**Unique Solutions Or Answers**."

Since the absence of a unique solution (which can only be provided by a true science) to a problem causes such havoc in any sphere of activity as discussed above, therefore a systematic method to solve problems forms, in no small measure, a vital area of any true scientific subject.

"*So long as we live among men, let us cherish humanity.* "

—Andre Gide (1869 - 1951)
French Writer, 1947 Nobel prize for literature

PART IV

The Realm of Postulates

Understanding Postulates

We need to define the term "postulate" at this stage of our work, because it will be used over and over throughout this book. By ***"Postulate"*** we mean, "***An assumption or assertion set forth and assumed to be true unconditionally and for all times without requiring proof, especially as a basis for reasoning or future scientific development.***"

Of course, if postulates are chosen as close as possible to the truth of the situation at hand, then they can be used as a basis for reasoning and, more importantly, as a solid foundation for a workable science.

On the other hand, totally arbitrary postulates can be formed, without any basis in truth or reality of the situation. Adoption of such arbitrary postulates would bring about a field of study, which is totally void of any valid truth or natural laws and would be a constantly shifting subject. The subject would be randomly workable and have a lot of false data. As an example, one may consider the field of art where the critic, the artist, the reviewer, and the public each have introduced their own arbitrary postulates to create a totally unworkable subject without the discovery of any natural laws. Based on this observation, the following conclusion can be drawn:

In any given subject, the number of arbitrary postulates made in a subject, which have no basis in fact or truth, inversely determine the workability of that subject. Moreover, the lesser the number of arbitrary postulates and the closer they are to truth, the more workable that subject becomes.

Thus we can see that in any given subject, the number of arbitrary postulates made in a subject, which have no basis in fact or truth, inversely determine the workability of that subject. Moreover, the lesser the number of arbitrary postulates and the closer they are to truth, the more workable that subject becomes.

"All riches have their origin in mind. Wealth is in ideas - not money."

—Robert Collier (1885-1950)
American Motivational Author

Physics and engineering have become two dominant fields in our highly technical society. There are a series of exact postulates, which have led to the current high degree of workability and successful application techniques and sophisticated technology.

The main postulates that have been discovered by observations of the physical universe can be summarized as:

1) *Existence of energy in many forms such as thermal, optical, electrical, magnetic, audio, electromagnetic, etc.*

2) *Existence of matter in gas, fluid, plasma, and solid forms.*

3) *Conservation of all forms of energy to hold valid at all times.*

4) *Conservation of any form of matter under all conditions and at all times (primarily used in classical physics).*

5) *Existence of a viewpoint to act as a reference point for all subsequent measurements. This postulate is implicit in physics.*

6) *Existence of a linear space in which to place matter and energy. This is another postulate implicit in physics.*

7) *Existence of a constant rate of motion or constant rate of change of particle's position in space. This postulate instantly leads to a linear time base, which is also implicit in physics and can be considered to be the seventh postulate. Such a constant and linear time base is*

> *currently supplied by the earth's spin around itself and rotation around the sun. It should be noted that this choice of time measurement is completely arbitrary but very convenient since the concept of time has existed from the beginning of man's civilization on earth.*

Time, in general, is an abstract consideration, which can be measured by any constant motion or oscillation and does not necessarily have to be in terms of seconds, minutes, hours, days, years, or even closely associated with the earth's movement in space. The fact that we take the notion of time for granted and measure it exclusively in terms of the above units is an insidious act of disregard for its more general concept and deeper connotations, of which more later.

In the above discussion, postulate #5 (viewpoint) is the most senior of the seven postulates, followed by postulate #6 (space), #1 (energy), #2 (matter), #3 (conservation of energy), #4 (conservation of matter) and #7 (time) in importance. However, the order of presentation of these postulates is, of course, to conform to the current state of physics texts, which make no mention of #5 and #6 as the senior postulates. However, the presented sequence of postulates is universally accepted as valid in the scientific community and most scientists take it for granted. This can be considered to be a major oversight and shortcoming in the extant physical sciences.

"The best ideas are common property."

—Seneca Wallace (5 - 65 AD)
Roman Philosopher

The Main Postulates

We can see that the physical universe, at first glance, appears to be built upon four *primary postulates* of:

a) **Space,**
b) **Energy,**
c) **Matter, and**
d) **Time.**

"Never, never rest contented with any circle of ideas, but always be certain that a wider one is still possible."

—Pearl Bailey (1918-1990)
American Entertainer

However, upon further examination of these four postulates we can see that the third and the fourth postulates have a common denominator, which allows us to further reduce the four primary postulates to only three. The new and reduced set of three postulates essential to the creation of the physical universe, are called the original postulates, which are but irreducible!
The three *original postulates* can be summarized as:

I. **Space**
II. **Energy (or Force)**
III. **Change**

Combining postulates (II) and (III), we can actually obtain the third and the fourth primary postulates discussed earlier (i.e., matter and time). For example, if we "change" the created space, in which certain energy particles exist, in such a way that the volume is reduced, then the energy particles become more condensed and thus we get "matter" (the third primary postulate).

On the other hand, if we "change" the position of a matter particle or an energy form in space, we get "time" as the fourth primary postulate. Upon these four primary postulates, which furnish the necessary building blocks of postulates for the creation of any universe or physical science, the microscopic and macroscopic universes are founded as shown below.

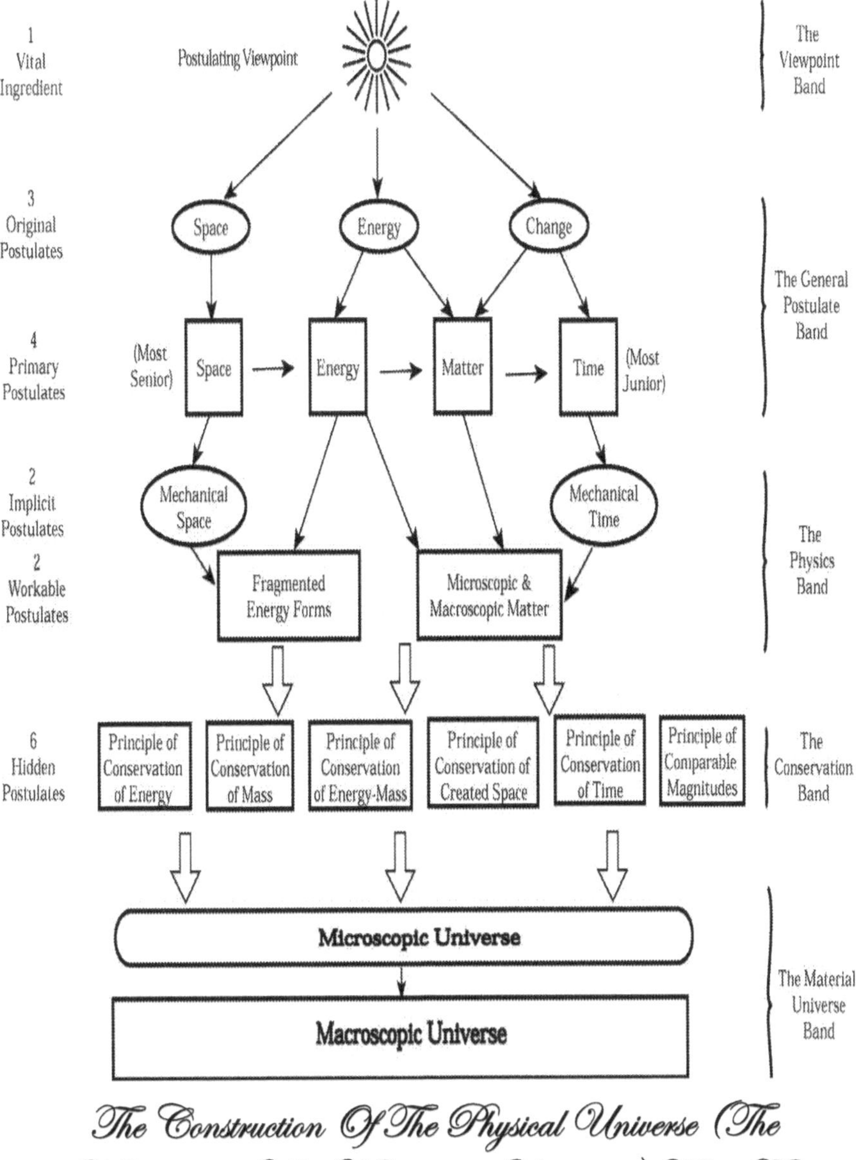

The Construction Of The Physical Universe (The Macroscopic And Microscopic Universes) From The Original And Primary Postulates.

Therefore, we can see that we only need three original postulates to create the entire physical universe—from an atom to a galaxy and beyond. The three original postulates bring a considerable simplification of the primary postulates and display a higher level of truth than one can ever obtain from a detailed examination of the components of the physical universe as done in the science of physics.

Of course, one should be reminded that once these three postulates, on a microscopic and macroscopic level, were implemented and put into effect on an automatic level of operation, the ensuing complexities of great magnitude, which was brought forth as a result may have well been outside the scope and grasp of the original postulates and may actually have been totally unforeseen to this day.

It is interesting to note that from pure observation, one can see that the third original postulate (change) has actually been put on automatic in the physical universe. We can take earth as an example. We can see that our planet revolves around the sun at a predetermined rate of 66,000 miles per hour while spinning around itself at a fantastic speed of about 1000 miles per hour—all done on an automatic, unthinking, and involuntary manner, causing an automatic time stream called mechanical time.

"The ability to convert ideas to things is the secret of outward success."

—Henry W. Beecher (1813-1887)
American Liberal Congregational Minister

Postulate Zero: The Concept of Viewpoint

"I cannot stop thinking that I died before I was born and that at my death I will return to the same state."

—George Christopher Lichtenberg (1742 - 1799)
German Physics Professor and Scientist

It can be observed that before the main postulates regarding the physical universe came into existence, postulate zero existed. We define this postulate as:

POSTULATE ZERO: *Is the existence of viewpoint as the source point of any universe.*

"One of the greatest pains to human nature is the pain of a new idea."

—Walter Bagehot (1826 - 1877)
British Political Analyst, Economist and Editor

The Scientific Postulates

Upon careful observation, it can be seen that the physical universe is constructed on a series of very exact postulates, which have been carried out to its final fruition. The end result of these series of postulates is a broad range of application mass, which includes such things as galaxies, star systems (e.g., solar system), planets (e.g., earth), compounds, molecules, atoms, electrons, so on and so forth.

The exact series of postulates however made, forms the heart of all of the extant man's knowledge about this universe. These postulates are implicit in all of our modern physical sciences, and have unfortunately never been heard of in subjects such as political science, art, social sciences, and the humanities. The postulates creating the physical universe form the backdrop against which all of our modern physical sciences are founded. Therefore, it certainly behooves us well to know completely what these postulates are and realize the role they play in our understanding of the current subject at hand.

For example, the workable postulates of physics along with the implicit and hidden postulates that have been mostly left out by almost all textbooks form the backbone of physics of which electricity is but a subset. The totality of these postulates led to the discovery of many laws and dynamic principles. Therefore, they are completely responsible for all past technical developments, present technological conditions, and all future applications and progress.

"Ideas can be life-changing. Sometimes all you need to open the door is just one more good idea."

—Jim Rohn (1930-)
American Motivational Speaker, Philosopher and Author

Any science is firmly built upon a handful of powerful postulates concerning the physical universe and several discovered principles about electricity, which with the help of the very exact and abstract science of mathematics, has brought about an enormous amount of application mass to increase our survival potential, and along with it has enriched our lives with a higher level of understanding about the planet and the solar system in which we dwell and call it our home. Moreover, the wave aspect of electricity has actually opened up the whole galaxy to our inquisitive minds and has made the center of our galaxy within our reach, to be able someday to explore it first hand.

The successful discovery and implementation of the postulates of a field of study such as "electricity" happens to be only one example of "how a universe can be constructed" such that it will persist and become useful to mankind.

It is interesting to note that even though postulates of physics are a cut-down and are actually different than the original postulates (or the primary postulates) that have gone into building the physical universe, it nevertheless exists as a workable subject within a certain range.

The one essential element that should never be forgotten in studying physics and engineering is the use of "viewpoint" (postulate zero) that has gone into the construction of theories of classical or quantum physics. This viewpoint is a composite viewpoint, which is evolved and polished through millennia of thinking men on the subject. These are the valuable final product of practical philosophers and scientists, who by using the scientific methodology as their stock in trade, have observed the material universe (the tangible) and its many natural phenomena and through trial and error, have honed their way to superior theory and eventual knowledge (the intangible).

"There is one thing stronger than all the armies in the world, and that is an idea whose time has come."

—Victor Hugo (1802-1885)
French Poet, Novelist and Dramatist

The founder's viewpoint, through thee ages, has led to the discovery of the very basic scientific postulates we have discussed in this book, which eventually won the physical universe many times over. These factors of a science should never be taken lightly, especially for one who strives to achieve a thorough understanding of the subject and wishes for deeper truth in all of the complex theories and abstract mathematics that has gone into describing physics and its allied subjects.

For example, the discovery of the basic postulates of the science of electricity from the beginning to end took a relatively short period of time compared to all other extant subjects. It took approximately a total of 300 years for it to be developed into a mature subject. In other words, from our vantage point at the present moment, and disregarding the total gap in the investigation of electricity from 600 BC to 1600 AD, it took only three centuries to progress from an almost zero of knowledge to a total knowledge and mastery of one of the most powerful subjects ever conceived on earth.

Over the last hundred years we have had nothing but applications of these very dynamic postulates, which were elegantly organized not only in conceptual description but also in mathematical form. Examples include radar, radio communication, RF/Microwave Integrated circuits (RFICs/MICs), space explorations, wireless applications, medical applications, surveillance, so on and so forth.

"***An idea that is developed and put into action is more important than an idea that exists only as an idea.***"

—Buddha, '*The Enlightened One*' (563-483 BC)
The Title of Indian Prince Gautama Siddhartha, Founder of Buddhism

Materialized Considerations

As one looks across the aggregate of modern physical sciences, one cannot help but notice that the bulk of these sciences are held up by considerations. These considerations are supported in part by a series of powerful postulates made by the founders and pioneers of the subject, which have shaped the entire subject as a living and growing entity. Of course, the postulates are furnished by a viewpoint or a composite viewpoint, and are developed through the many centuries that physical sciences have formally existed.

These postulates are capable of changing the existing state of a science and/or creating new areas of study. In actual fact, the postulates trickle down through natural laws, theorems, design techniques, charts, and tables to the final byproduct which is the application mass.

Conversely, one can follow this line of logic and observe that before an application mass can exist there must be a manufacturer, before that a designer, and before that a postulator of that application mass; and before all of these, there *comes a viewpoint that is existing exterior to this process and is solely concerned with observation and postulation of new and orderly thoughts, which reflects nature and its inner workings.*

"Everything begins with an idea."

—Earl Nightingale (1921-1989)
American Motivational Writer and Author

Thus, beyond the physical entity of a product or an object lies a higher entity called a viewpoint that through a mental process arrives at the postulates, and eventually makes it possible to create the application mass. The key postulates to be made in any science starts with some basic postulates concerning space, energy, matter and time. These are first stated purely as concepts, which then materialize as forms as shown below. These postulates, even though initially made on a sequential basis, are implemented simultaneously in actual practice. The dual of the postulating viewpoint is the observing viewpoint, which is also shown in the Figure below.

"The key to growth is the introduction of higher dimensions of consciousness into our awareness."

—Lao Tzu (600 -531 BC)
A Chinese Philosopher

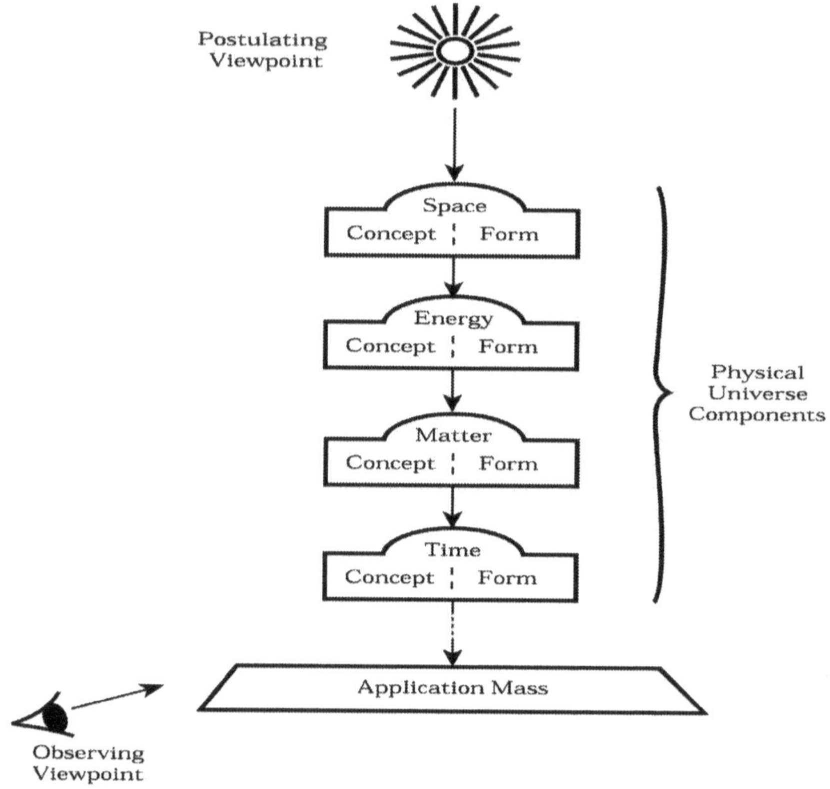

The Relationship Amongst The Postulating Viewpoint, The Physical Universe And The Observing Viewpoint.

⊶☙✶❁✶❧⊶

The Postulate's Byproducts

Moreover, any extant application mass carries with it a physical stamp of the simultaneous implementation of the basic postulates and has all of them built into its constituent components at each moment in time.

So one can see that beyond the physical universe lies thought in its purest form: an "analytical and aware thought" that has the ability to create postulates, form concepts, create space, create energy, matter, and put them in motion as an application mass and thus create a time stream as shown below.

"A man is not idle because he is absorbed in thought. There is a visible labor and an invisible labor."

—Victor Hugo (1802-1885)
French Poet, Novelist and Dramatist

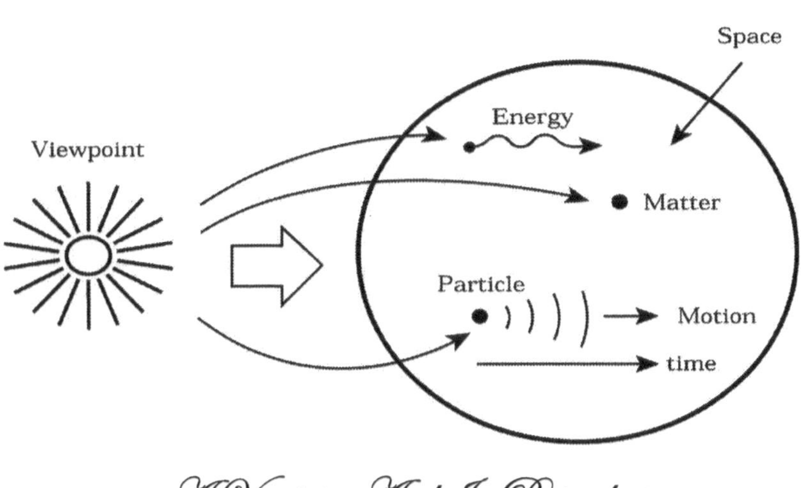

A Viewpoint And Its Byproducts.

A Summary of the Postulates

There are a handful of basic postulates from which the bulk of the materials in the physical universe can be derived systematically. These postulates can be roughly summed up as:

I. SPACE- Let there exist an enormous space spanning billions of light years in dimension

II. ENERGY- Let there exist very light, fast-moving and random energy particles, which are in constant motion, high agitation and total confusion.

III. CHANGE- Let these energy particles be constantly changing on an automatic and yet random basis. The "change postulate" brings forth the concepts of matter and time. These two sub-products of the "change" postulate are important aspect of the physical universe.

IIIa. Matter- Condensation of energy particles (i.e. shrinkage of the created space), which leads to the concept of atoms and molecules on a microscopic level and objects on a macroscopic scale; and,

IIIb. TIME- The concept of change of particle position or change of location in space, leads to the concept of mechanical time. Thus, we can see that the change postulate leads to a new postulate called the "time postulate," which brings forth a time stream for every region of space where there is energy.

Diagram below shows how the viewpoint through the thought universe creates the physical universe.

"I want to know God's thoughts. The rest is just details."

—Albert Einstein (1879-1955)
German Born American Physicist, Nobel Prize in 1921

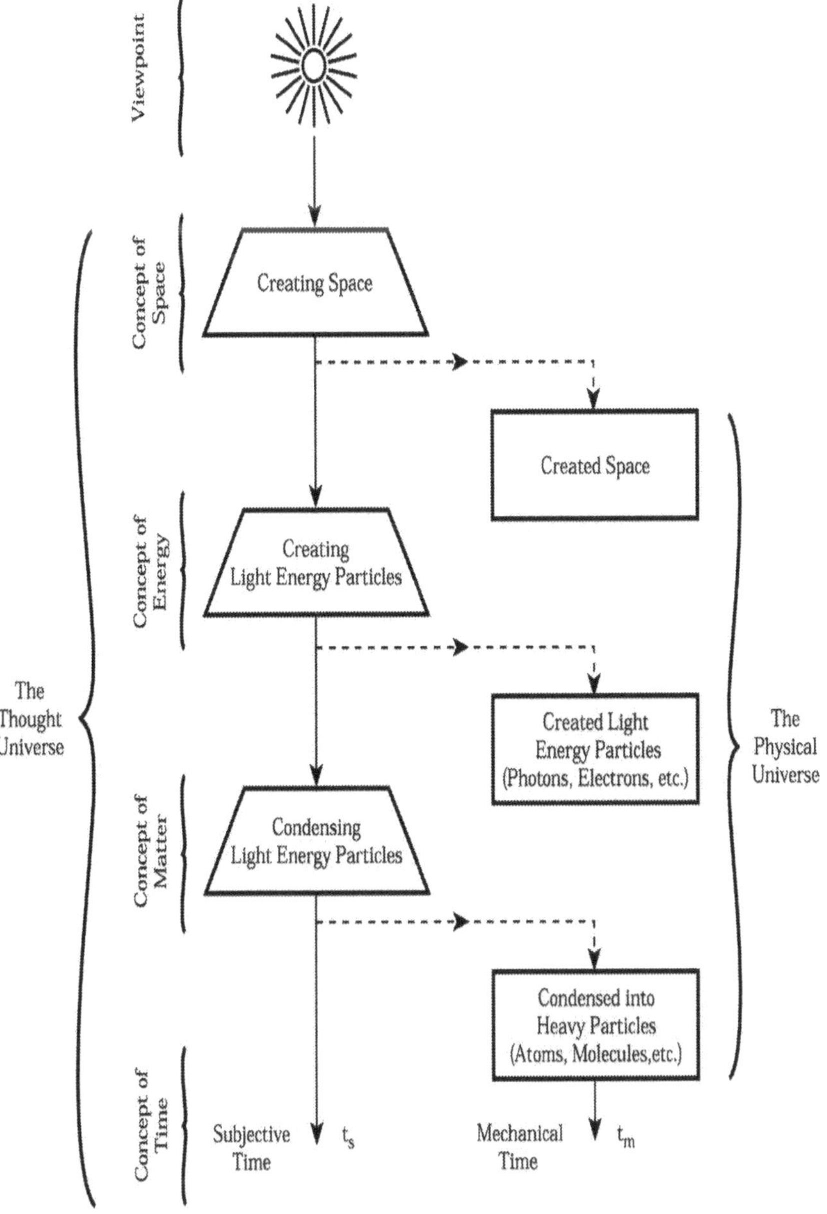

A Diagram Showing How The Thought Universe Creates The Physical Universe.

Workability of a Scientific Postulate

The degree of workability of a postulate is established by three factors:

a) *The degree it explains an existing phenomenon,*
b) *The degree it predicts new phenomena hitherto unknown, and*
c) *The degree that it does not require arbitrary data to be called into existence for its explanation.*

The above three factors form a powerful set of criteria, against which any scientific premise or, in general, any proposed data can be tested for its potential worth with tremendous success.

For example, the field of "arithmetic" has been with Man for a very long time. The founding postulates of this subject are: a) Addition, b) Subtraction, c) Multiplication, and d) Division.

These postulates were uniformly adopted at some time in the past by the founder(s) of the subject and are assumed to be true unconditionally and at all times and thus, cannot be proven.

The above series of postulates in arithmetic, based on the decimal number system, have created a workable subject, which is very valuable to many fields. But the moment we deviate into another number system such as "the binary number system" as used in "digital logic," these postulates become unworkable. They have to be replaced by the three workable postulates of "Boolean logic," which are AND, OR, and NOT.

"There is a single light of science, and to brighten it anywhere is to brighten it everywhere."

—Isaac Asimov (1920-1992)
Russian Born American Science-Fiction Writer and Biochemist

Scientific Axioms

There are many axioms that can be derived from a handful of postulates, which form the foundation of any science. We can define an axiom as, "*A statement universally accepted without proof. In other words, we mean a self-evident truth, which is agreed upon without rigorous tests or trials.*"

Any workable subject has a series of axioms built into it as an essential part and parcel of its existence. These axioms express a series of valid observations, which are true without proof.

"The most dangerous of all falsehoods is a slightly distorted truth."

—George Christopher Lichtenberg (1742 - 1799)
German Physics Professor and Scientist

There is much material that can be derived from a basic set of postulates and axioms. The derived material, if based upon the actual postulates and axioms, can be shown to be also valid. However, a total neglect combined with a complete unawareness or even misunderstanding of the basics of a subject will produce materials, which will contain inaccurate and misleading information.

One must be well equipped to handle inaccurate sources of information, if one is shooting for a long-term survival in the scientific arena and plans to make a successful study of it.

"It has long been an axiom of mine that the little things are infinitely the most important."

—Arthur Conan Doyle, Sr. (1859-1930)
Scottish Writer, Creator of the Detective Series, Sherlock Holmes

The Implicit Postulates of physics

"Thoughts are the seed of action."

—Ralph Waldo Emerson (1803-1882)
American Essayist

There are two postulates that are implied in Physics but never asserted or discussed openly. These two are:

a) *The existence of a continuous and linear created space, in which all of the energy and matter particles can be placed, and*

b) The existence of a continuous and linear index of motion commonly known as mechanical time, measured in seconds, minutes, hours, etc.

"The important thing in science is not so much to obtain new facts as to discover new ways of thinking about them."

— William Bragg, Sr. (1862-1942)
British Physicist

The Hidden Postulates of Physics

"Three things cannot be long hidden: the sun, the moon, and the truth."

— Buddha, '*The Enlightened One*' (563-483 BC)
The Title of Indian Prince Gautama Siddhartha, Founder of Buddhism

We can explore the subject of hidden postulates and realize that the biggest hidden postulate is, "The existence of viewpoint," which can be labeled as postulate zero. However, there are two other hidden postulates, which have been omitted in most texts and yet they rule the world of physics and engineering. These are:

a) The principle of conservation of energy, which is held valid under all conditions and for all times.

b) The principle of conservation of mass, which is held valid under all conditions and for all times.

"Simplicity is the ultimate sophistication."

—Leonardo da Vinci (1452-1519)
Italian Painter, Sculptor, Architect and Engineer,
whose Genius Epitomized the Renaissance Humanist Ideal

The Common Denominator of Postulates

"The common curse of mankind, -- folly and ignorance."

—William Shakespeare (1564 - 1616)
English Dramatist, Playwright and Poet

Excluding space, we can see that the common denominator of all of the postulates concerning the physical universe can be summed up into two extremely powerful postulates:

a) *Energy or force, and*
b) *Never-ending Change.*

"Common sense is not so common."

— Voltaire (1694-1778)
French Philosopher

These two common denominator postulates are interwoven throughout the entire physical universe and have made it a universe, which is difficult to grasp with full certainty, because:

a) Under the heading of force, we have the general class of forces, such as impacts, impulse, mechanical force, electromagnetic force, gravitational force, chemical force, etc. This category also includes energy, such as EM waves, sound waves, electric fields, magnetic fields, electromagnetic radiation, X-ray, etc.

b) In the category of never-ending change, we have change of particle's position in space randomly at the subatomic level, which is unpredictable and uncertain; change of an object's position, which is predictable with a high degree of certainty on the macroscopic scale—an apparency and not actuality.

The most common factor associated with change is time. Time is the main index of measurement of change and, strictly speaking, is an abstract concept based on the considerations of the viewpoint. It is a "pseudo-commodity" since it cannot be exchanged between two parties, like the other three components of the physical universe: matter, energy, or created space.

"Believe nothing just because a so-called wise person said it. Believe nothing just because a belief is generally held. Believe nothing just because it is said in ancient books. Believe nothing just because it is said to be of divine origin. Believe nothing just because someone else believes it. Believe only what you yourself test and judge to be true."

— Buddha, '*The Enlightened One*' (563-483 BC)
The Title of Indian Prince Gautama Siddhartha, Founder of Buddhism

Role of Postulates in Life

As a side comment and an analogy to what we have been discussing, we can observe that one's own life appears to be a universe of created things. The viewpoint furnished by the person himself is constantly in a state of postulation of future efforts, future goals, and future personalized application mass (P.A.M.), which we may call personal possessions, for short.

These postulates, as they are arbitrarily set up to be true in one's own universe, when they have no basis in truth and actuality of one's condition in one's life, usually end up opposing each other and thus, bring about a very confusing universe of created things. This particular state of affairs does not lead to the creation of much future personalized application mass and, in fact, may cause a reduction or a loss of the present ones. In other words, postulates lead to methods of survival in life and to certain physical possessions, including one's body and its many characteristics as shown below.

"Resolve and thou art free."

—Henry W. Longfellow (1807-1882)
The Most Popular American Poet in the 19th century

The application mass of the science of physics consists of devices, machines, etc. In a similar fashion, we can see that one's physical possessions, or the physical body and its characteristics (such as its shape, hairdo, health, etc.), where the body can be considered to be analogous to a machine, are actually the personalized application mass of the viewpoint, furnished by the life force.

The only difference between these sets of postulates and the ones used in physics or any other physical science in general, is that the scientific postulates are few and seldom change and have a deep basis in truth regarding the physical universe, whereas one's own postulates are many, change with time, and usually concern a goal or

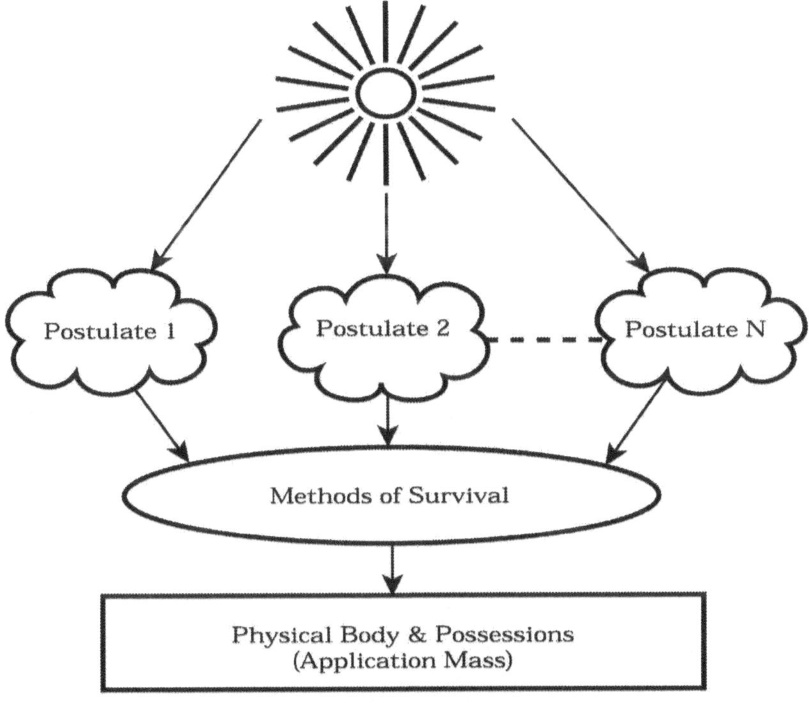

From Viewpoint To Postulates To Methods Of Survival And Eventually Arriving At The Physical Body And Possessions.

a wish, which has very little to do with the current conditions in the physical universe.

Furthermore, one can see that since postulates make the space of a universe and since one's postulates shift in time, then one's space changes as time progresses.

"Life is rather like a tin of sardines - we're all of us looking for the key."

—Alan Bennett (1934-)
English Stage and Film Actor and Writer

The Modern Scientist

By observation, we can conclude that modern Man, including all scientists and engineers, are below the level of postulates of sciences. That is to say, our role as scientists in all of this has been primarily to find out the operating principles behind an observed phenomenon and discover a number of basic facts and consolidate these findings into a series of laws, techniques and rules of thumb, which eventually feed a great many inventions and application mass to be used by the rest of humanity.

"The scientist is not a person who gives the right answers; he is one who asks the right questions."

—Claude Levi-Strauss (1908-)
French Philosopher

It is very easy to visualize the process of building a house, where it all starts with a dreamer (a postulator) and his dream (a postulate). His dream goes through a very common process of construction with which we are all too familiar. It starts with an architect, who takes that dream and turns it into the space of the house along with all the required hardware and materials all neatly designated on paper, i.e., he creates the space, energy, and mass of the house according to the building codes which are a series of pre-designated agreements intended to make the final product safe and habitable. Next, with proper city permits, the building materials are brought in, and finally the house, which is the application mass of the postulate, materializes.

It is rather easy to see that the original postulate was the cause point behind the reality of the house. At the time of postulation of the house, the postulator was at total cause and could change any and all parameters of the house at will. In other words, the postulates are flexible up to the point of execution of plans. From the moment of construction and onward, none of the postulates could be changed and the postulator would suddenly find himself the effect of his own causation.

"Scientists dream about doing great things. Engineers do them."

James A. Michener (1907-1997)
American Novelist and Short-Story Writer

Such is the case with the sciences and their basic postulates. At the present moment, it can be said that Man is the effect of any and all scientific postulates currently in force in this corner of galaxy. Man's role has been: a) To establish and derive the workable laws, and b) To utilize these laws effectively and build application mass in order to enhance his own survival potential on this planet and be able to export his technology to colonize outer space.

There has been a considerable failure on Man's part to rediscover the basic postulates made by a viewpoint (or viewpoints) at a distant past and approach the study of the physical universe from this angle.

Man, being a relatively new creature, has emerged as the ruling party primarily due to his inherent abilities in reasoning communication, observation, analysis, etc., which have brought him much success in enhancing the survival potential of his own species and his symbionts.

"The scientists split the atom; now the atom is splitting us."

—Quentin Reynolds (1902-1965)
American Correspondent, Writer and Editor

It could be further stated that in his current state of development, Man is in a high state of admiration (or almost worship) of any application mass (later referred to as generalized application mass) commonly referred to as, "the physical universe." He looks at it with a certain level of awe and fascination. The highest level of understanding he has achieved through his observations of the physical universe is the level of "workable laws" such as Maxwell's electromagnetic equations (a series of four laws), Newton's laws of motions (a series of three laws), etc.

To get a deeper truth and understanding about the subject, one needs to ascend to the level of postulates in order to face the ultimate point upon which the reality of the physical universe is founded.

"Each morning puts a man on trial and each evening passes judgment"

—Roy L. Smith (circa early 1900s)
American Author

The average scientist's awareness is well below the level of existence of postulates and their influence as the main driving force behind our universe. This has been the primary factor, which has necessitated the author to make this commentary and to point out what is propelling the physical universe into existence—it is not the raw force but the postulates behind it! This is a rather gross oversight by the scientific community. In other words, for a scientist to mistake the application mass for actuality and totally neglect the postulate as the main proponent of its existence is rather unforgivable, now that we know the power of postulates in sciences.

The rest of the items, such as laws and axioms are all derivable from the basic postulates but have been underemphasized; whereas, rules, techniques, symbols, and mathematical operations have been unnecessarily overemphasized by our current educational systems and the engineers and scientists at work.

By observation and close examination of the facts, we note that the cause-effect relationship between the physical universe and the postulates (originating from the thought universe) creating it, has been lost in the sciences and they have been turned into the "study of effects."

In other words, modern sciences have taken on the study of created spaces, created energies, created masses, and the resultant created time periods. Our "dyed-in-the wool scientist" has assumed the role of the observer and the reporter of effects.

"Learn from yesterday, live for today, hope for tomorrow. The important thing is not to stop questioning."

—Albert Einstein (1879-1955)
German Born American Physicist, Nobel Prize in 1921

PART V

The Phenomena of Space

Understanding Space

Space is a consideration, which is implicit but actually not embraced in the science of physics, since it is one step behind the existence of matter and energy and needs to exist first, before matter and energy can exist.

The study of the existence of matter and energy as done in physics, presupposes the existence of a linear created space in which we are placing these entities.

Linear space is the first consideration behind the science of physics, which is implicitly alluded to but not directly addressed. Thus, it is essential that we regress one step back from energy and matter and examine space in order to build the science of physics from the ground up—metaphorically speaking!

We can define **space (Also Called Mechanical Space or Created Space)** as: *the delineated and designated region within a closed boundary line or a closed and bounded surface within which all things under consideration can be placed as shown on the next page.*

Using this definition we can see that mechanical space, in simple terms, is the continuous expanse extending in all directions within which all things "*under consideration*" exist.

We loosely refer to "mechanical space" or "created space" as "space" in most of this text and in all physics texts. But, it gradually becomes obvious that the source of space lies with the "viewpoint," since it is the viewpoint that delineates a "boundary line or boundary surface" and within it designates a region called "space."

"For the wise man looks into space and he knows there are no limited dimensions."

—Lao Tzu (600 -531 BC)
A Chinese Philosopher

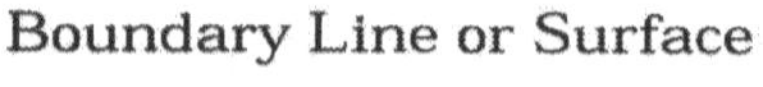

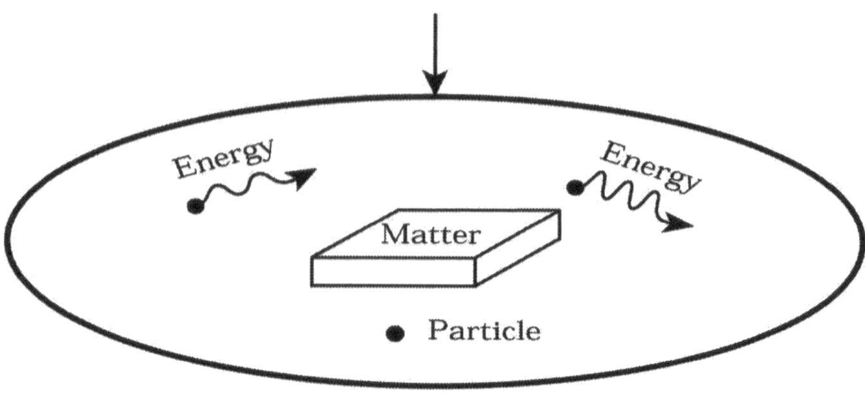

The Definition Of Space, Or More Precisely, A Created Space.

Once this has taken place, then the resultant space is usually referred to as "created space." Without this action of the viewpoint, we do not really have a "created space" to work with. Moreover, from the position of the viewpoint, there are infinite number of lines extending to the boundary line or surface, where each such line is called "a dimension."

It should be further noted that when we are analyzing a problem, we are dealing with a created space, whereas when we are involved in the design process (whether analytically or with the help of mathematics, software, etc.), we are creating space in which we shall place the final designed product. Therefore, we can see that engineering, in actuality, is an interplay between the analysis of the energies and masses existing in a "created space" of a problem or "creating space" in order to place the energies and masses of the future designed products and forms.

"Life, forever dying to be born afresh, forever young and eager, will presently stand upon this earth as upon a footstool, and stretch out its realm amidst the stars."

—H. G. Wells (1866 - 1946)
English Novelist, and Historian

The Space Age

The door to space explorations has gradually opened up and a new age has been ushered in. This new age, to which we are just standing at its threshold, is the age of space opera. This new age has been preceded by many ages. Important amongst them are the age of communication, age of industrial revolution, age of electronics, age of computers, and the age of information.

The age of space opera is different than the "***Space Age,***" which is the period starting with the launching of the artificial satellites and manned vehicles (October 4, 1957).

The age of space opera is the age of interplanetary travel and migration to other planets in the galaxy using space vehicles having speeds greater than anything possible on earth. Such high speed vehicles are a must in a space opera society since one would then be able to travel large distances in a short period of time.

The reality of this space opera era is gradually dawning on us. Its feasibility, at first, may appear to depend on high-speed space vehicles, but upon close examination, we find that success of such space travels depends on the transmission and reception of radio frequency waves, which would be invaluable for navigation and processing of information about the internal or external conditions.

"The Earth is just too small and fragile a basket for the human race to keep all its eggs in."

—Robert Heinlein (1907-1988)
American Science-Fiction Writer

Reference Frames

Considering the concept of "created space" and its relation to the "viewpoint," we can see that that there are an infinite number of dimensions, extending in all directions from the viewpoint to the boundary surface.

However, in the sciences, this situation has been greatly simplified through the use of a *reference frame* (also called a frame of reference), *which is a coordinate system for the purpose of assigning "positions and times" to events.*

A reference frame consists of a reference point and a set of three orthogonal (or perpendicular) graduated axes. For example, in the Cartesian coordinate system, we have three perpendicular axes for each of the principal dimensions (length, width, and depth) to bring about a "coordinate system of measurement" as shown below.

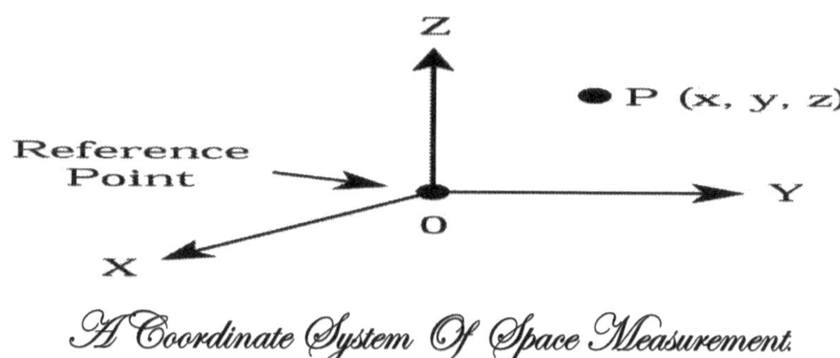

A Coordinate System Of Space Measurement.

Observation: *Utilizing orthogonal coordinate systems (such as Cartesian, Cylindrical, etc.) in the analysis of engineering problems, it has been found that it is only necessary to select three main dimensions out of an infinity of dimensions, in order to identify any given point in space uniquely. Thus, the use of the uniqueness axiom has been instrumental in developing the present workable reference systems.*

Reference frames are essentially an organizing tool set up to measure a created space. Reference frames organize the created space and through the measurement system that they employ, one may consider them a method of bookkeeping of that space. Their main functions are:

a) To provide orientation,
b) To designate a reference point, and
c) To identify each point uniquely in the three dimensional created space.

The use of a coordinate system as a standard and indispensable tool in all of scientific analysis is based upon the fact that any observation point can be uniquely located in space once all of the three coordinates of that point are specified based upon the "Uniqueness Axiom."

We usually note that the concept of space in engineering and physics brings about the idea of a coordinate system. The concept of a coordinate system exists in all aspects of physical sciences and mathematics. Mathematics attempts to provide ideal solutions to abstract or idealized problems posed by a science. These problems involve space and other variables, to which mathematics is providing answers, through abstraction and symbolization of physical entities.

The use of a coordinate system interwoven throughout the physical and engineering sciences has been made to be so inevitable and essential to the understanding and analysis of problems and design of new structures that without it, most "sciences" become unworkable and highly speculative.

Thus, as can be seen from the above observations, space assumes a senior position in construction of any universe (particularly the physical universe) since it must exist first before other entities such as matter and energy can be located within it.

"Don't agonize. Organize."

—Florynce R. Kennedy (1916-)
American Lawyer, Activist, Speaker, and Author

Curved Space vs. Linear Space

It should be noted that under heavy force fields (such as strong gravitational fields, electric, or magnetic fields), space can appear to be curvilinear, which means that the shortest path between two points is along a curved line and not a straight line. The notion of curved or nonlinear space is an apparency and not an actuality.

The actuality is that the strong fields bend the flow lines and give the illusion of nonlinear space. For example, light from a heavy gravity star (such as our sun) reaching a planet with a relatively high degree of gravitational field (such as earth) will bend while leaving the star and bend again upon arriving at the planet. This will give the illusion that the space, at these speeds, is curved.

Another example is the light reaching earth from very far away stars, where their emitted light to reach earth must pass by a nearby galaxy. In such a case, light reaching earth gives an apparent source, which is a false location for the star due to the bending of light as it passes by the heavy-gravity galaxy, as shown below.

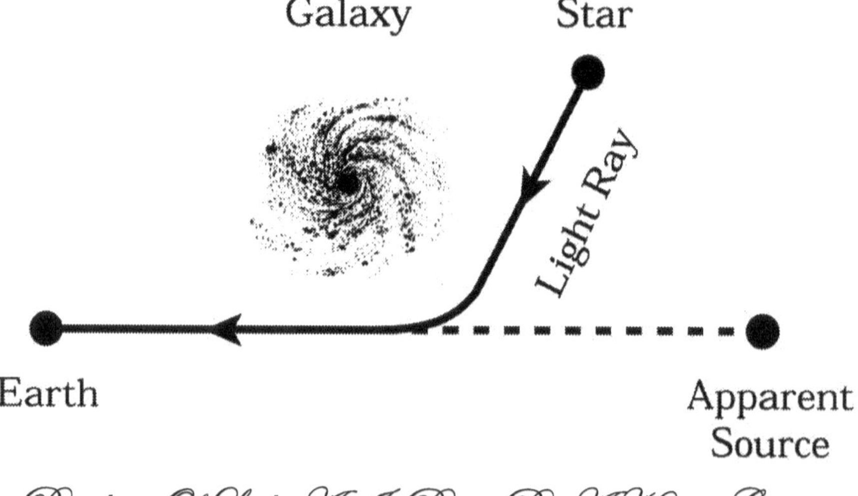

Bending Of Light As It Passes By A Heavy-Gravity Galaxy.

The created space remains linear; it is just curved field lines under heavy forces that give it a curved quality or characteristic. In such a case, the shortest path between two points is no longer a straight path but a curved line.

As a further proof of the linearity and non-curved-ness of space, let us consider our earth in its present condition. From our earlier discussion, we know that our Milky Way Galaxy containing billions of stars, planets, black holes, etc. is rotating around its axis.

Therefore, it is highly probable that the linear space that earth is currently occupying, belonged to an extremely dense star such as a black hole millions of years ago. According to the "curved-space theory," the space near a black hole is highly nonlinear, thus, we can see that at the present moment the nonlinearity has abruptly disappeared.

In other words, the apparent nonlinearity and "curved-ness" of space has turned into a "non-curved-ness," which makes us ask a very fundamental question: "Is it the space that is curved or is it the fields and condensed energy forms occupying that space, which give it a nonlinear or curved quality?" With the above preamble, the answer is very clear cut.

To understand this illusion further, one needs to visualize oneself exterior to the interplanetary space and observe the whole event of "light travel" from this outside vantage point. From this exterior point one shall see a created space spanning billions of light years filled with energy particles and under the influence of many force fields. It is not the space but what is in it that makes it behave in a nonlinear fashion. The illusion will cease and the linearity of the space will be revealed.

"The earth is the cradle of humankind, but one cannot live in the cradle forever."

—Konstantin Tsiolkovsky (1857-1935)
Russian Visionary and Pioneer of Space Travel

Space Warps

Due to the fact that created space is linear, concepts such as space warps and other space anomalies seem to be a figment of the imagination of science fiction writers and movie producers. The concept of space warp, wherein the created space has a discontinuity in it or acts nonlinearly, can safely be discarded as purely a theoretical concept at best.

Space warp as a concept can be easily expressed mathematically, and thus, one can work out all of its properties in terms of the characteristics of the warp. For example, one can set up a logarithmic space wherein successive equal distances in a created space have a value ten times of the previous equidistance marking. This space appears to start on a small scale and gradually rises by powers of ten to have a large value as higher demarcation marks are encountered. This means that, for instance, if one travels 10 miles in the first interval of space, in the second equal interval he would travel 100 miles and in the third interval 1000 miles and so on—highly nonlinear space.

"Hope warps judgment in council, but quickens energy in action."

Edward G. Bulwer-Lytton (1803-1873)
British politician, Poet, Critic and Prolific Novelist

One could also set up a discontinuity in space, where half of a created space is linear and the other half is logarithmic space. The interface between the two half regions is the location of the discontinuity or singularity, as it is called in mathematics.

These are workable concepts in mathematics, which is a highly abstract and idealized subject. The reality of our physical universe is quite something else.

Such a concept, as entertaining and amusing as it may be to science fiction writers, film studios, and movie fans, does not happen to be true and is not in the make-up of the physical universe.

Truth be told, even if such a thing as a space warp exists, it is not the property of space but of the energy fields occupying that space. It is the lines of force of energy in that space that color that region of space and make it behave nonlinearly! So let us recognize the source of the nonlinearity! This fact makes us realize that it is not the innate property of space itself but of what energy field occupies that space that would create a particular space anomaly. It is an apparency and not an actuality!

The proof is when the energy fields have been removed from a region of created space, it acts once again linearly as it was originally set up to be. As a valid proof in support of the concept of linearity of space, we can make a few observations as follows:

a) From intimate contact with the immediate space surrounding one on earth, one can conclude that created space acts linearly, and no space warp or nonlinearity exists on earth today.

b) From deep space observation, it is a well documented fact that the solar system, as well as the Milky Way galaxy as a whole, is rotating about its axis, completing one revolution once every 200 million years (i.e., the time period of rotation $T=2x10^8$ years). Therefore, the space that the earth currently occupies could have been easily occupied by another star or a black hole of enormous density and extremely large gravitational field, millions of years ago. Such a large gravitational field causes a highly nonlinear type of space to be set up, where light bends and completely travels back to the planet and thus, does not escape the surface.

c) Now, if space was nonlinear innately, then when earth would occupy such a highly nonlinear created space, we would have nonlinear space in our midst. That is to say, in such a created space, light no longer would travel on a straight line as we observe it on earth today.

d) Moreover, the fact that earth is occupying these once nonlinear spaces at this moment and our first-hand observation tells us that they no longer behave nonlinearly, makes us examine the concept of space more closely and make a basic conclusion about the nature of space:

In conclusion we can see that *created space is innately linear but can be made to appear or behave nonlinearly by strong energy fields.*

"No matter how vast, how total, the failure of man here on earth, the work of man will be resumed elsewhere. War leaders talk of resuming operations on this front and that, but man's front embraces the whole universe."

—Henry Miller (1891-1980)
American Author and Writer

PART VI

The Nature of Energy

Types of Energy

The concept of energy is junior to that of space and deals with the concept of force—the universal force that the ancient philosophers have talked about. The concept of energy in the physical universe takes on many forms. Energy is encountered in all aspects of existence, particularly in the scientific arena.

Energy is one of the foremost entities, which is defined as, " *The capacity or ability of a body to perform work.*" This definition can be applied to particles and objects in both quantum and classical physics.

"When you are in the valley, keep your goal firmly in view and you will get the renewed energy to continue the climb."

—Denis Waitley (1933-)
American Motivational Speaker and Author of Self-Help Books

Furthermore, energy of a particle or object can be subdivided into two parts:

a) Potential energy, due to its potential motion, and
b) Kinetic energy, due to its actual motion.

By "potential energy," we mean "unrealized or unreleased motion" type of energy, whereas by "kinetic energy," we mean energy due to "actual motion."

For example, a traveling rocket has both a potential energy (due to its height above ground) and a kinetic energy (due to its motion). As another example, an electron moving with velocity (v) having a distance (d) above a large positively charged plate has both kinetic and potential energy as shown on the next page.

"Our universe is a sea of energy—free, clean energy. It is all out there waiting for us to set sail upon it."

—Robert Adams (1921-)

New Zealandic Electrical Engineer, Inventor of a motor /generator converting the perpetual motion of sub-atomic particles into electric power.

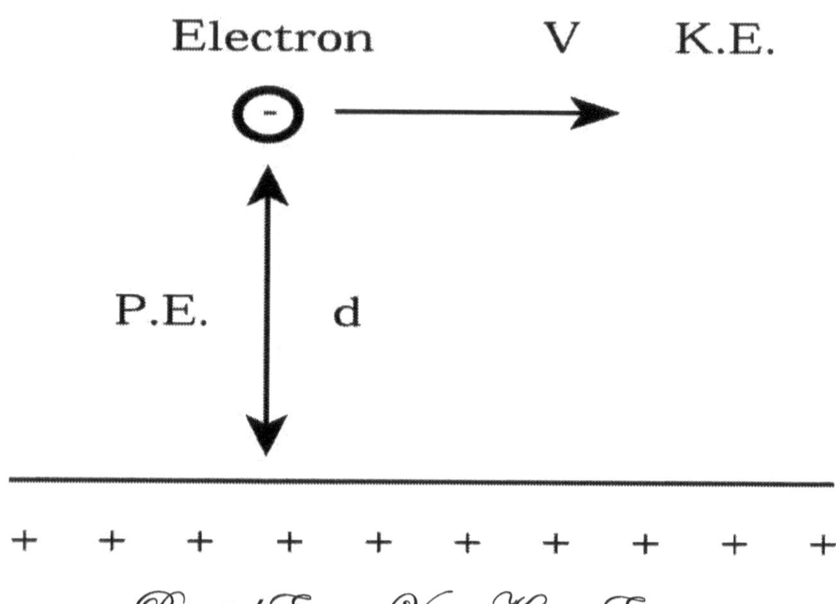

Potential Energy Versus Kinetic Energy.

⊶ ⁂ ⊶

Potential Energy

Potential energy (P.E.) is defined to be "*Any form of stored energy, which has the capability of performing work when released.*"

This motion is brought about because of the proximity of particles relative to one another. Therefore, this type of energy could also be called "proximity energy."

For example, the opposite charges in the two terminals of an electric outlet are in close proximity to each other and thus, have a stored potential energy. However, when these two opposite charges are allowed to move toward each other (such as through a piece of resistive wire), the release of potential energy first translates into a quantity of force on the electrons followed by an energy transfer to the wire in the form of electric power.

This means that the electrons moving in the wire over a period of time perform work (against the resistance of the wire), which eventually converts it into heat in the resistive wire. This process in simple terms could be considered to be the conversion of potential energy into kinetic energy.

An example of the reverse process is a power plant where coal or gasoline is burned to turn the motor of the electric generator, where the electrons flow to one terminal and away from the other terminal of the generator (kinetic energy) to set up an electric field between the two terminals, which would be transported by transmission line for consumption as unreleased motion, or potential energy.

"Continuous effort - not strength or intelligence - is the key to unlocking our potential."

—Winston Churchill (1874-1965)
British Orator, Author and Prime Minister During World War II

Another example would be a battery. Energy stored in the two terminals of the battery is potential energy until released as kinetic energy in the form of an electric current, which translates into force as an intermediary agent to perform work such as turning on an engine, etc.

"Free energy does not mean perpetual motion. Extracting power from the vast sea of energy that surrounds us is no more perpetual motion than a photovoltaic cell. It is a conversion of energy from one form into another form that is usable -- not the creation of energy out of nothing."

—Sterling D. Allan (circa 1965-)
American Founder of Organizations for Alternative Energy Sources

Kinetic Energy

Kinetic energy is defined to be, "The energy of a particle or an object due to its motion."

As an example, a moving electron in a conductor has a kinetic energy, which could be converted into another form of energy such as in an incandescent filament to give off light or could give up its kinetic energy to a highly resistive medium and create heat, as in an electric oven.

"All motion is cyclic. It circulates to the limits of its possibilities and then returns to its starting point."

—Robert Collier (1885-1950)
American Motivational Author

Classical physics (also called non-quantized or continuum physics) is the branch of physics based on concepts established before quantum physics, and includes materials in conformity with Newton's mechanics and Maxwell's electromagnetic theory.

In this branch of physics, distribution of energy (or matter and other physical quantities) is studied under circumstances where their discrete or quantum nature is unimportant, and they may be regarded as continuous functions of position.

On the other hand, quantum physics (also called quantum mechanics or quantum theory) is the study of atomic structure, which states that an atom or molecule does not radiate or absorb energy continuously. Rather, it does so in a series of steps, each step being the emission or absorption of an amount of energy packet (E) called a quantum. It differs from classical physics in that it generalizes and supersedes it, mainly in the realm of atomic and subatomic phenomena.

Kinetic energy is a commodity, which is encountered in all aspects of physics, whether quantum or classical. However, its definition and concept is different in each branch, which brings us to subdivide it into two categories of motion, small motion (such as an electron in an atom) and large motion (such as a traveling arrow in the air).

"All peoples everywhere should have free energy sources."

—Nikola Tesla (1856-1943)
Croatia Born American Inventor and Engineer

Electric Charge

The rock bottom foundation of the whole subject of electricity is electric charge, which is a form of energy. Charges in motion create electric current (a flow) between two points while non-flowing charges lead to static electricity and charge storage concepts.

"At electric speed, all forms are pushed to the limits of their potential."

Marshall McLuhan (1911-1980)
Canadian Educator, Writer and Social Reformer

To continue our study in a methodical fashion, it is imperative that we define one of the most misunderstood technical terms of all times and that is "electric charge," defined for the small and large scale of observation as follows.

I. (On a Small Scale) **Charge** *of elementary particles of matter (such as electrons and protons),* is *a basic property (or an essential attribute or quality) which is capable of creating a force field in its vicinity. This built-in force field is a result of stored electrical energy*, and

II. (On a Large Scale) **Charge** *of an object (such as a battery or a generator), is the net or algebraic sum of the charges of its component parts and may be zero, a positive, or a negative number.*

Understanding these two aspects of charge is a milestone in one's grasp of one of the most exciting and vital forms of energy: electricity, more of which later.

"***Static charge really means separated opposite charges.***"

—William J. Beaty (1961-)
American Electrical Engineer

Electricity

There are two forms of electricity: **Static and Kinetic**.

Static electricity *is that form of electricity which is not in motion, such as the charges resident on the plates of a capacitor.*

Kinetic electricity *is that form of electricity that is caused by a potential difference between two points and can be expressed in terms of the motion of charges in a created space between those two points such as an electric current in a wire or charged ions moving in an electrolyte.*

Static electricity is separated opposite charges, and does not involve any electric currents or magnetic fields. Therefore, it has very limited applications in actual practice.

On the other hand, in order to have *kinetic electricity*, we need to have two factors present:
a) The prime mover, and
b) The moved.

The "prime mover" is the electric pressure (also called the electromotive force, emf), measured in volts, that causes an electric current to flow in a circuit; the "moved" is the electric charge.

The concept of charge in motion brings into view the practical idea of electric current on a macroscopic level defined as, "*The rate of net transfer of electric charge past a specified point or across a surface.*" Electric current is measured in Ampere and has a short-hand definition: *"Charge per unit time."*

"On matters of style, swim with the current, on matters of principle, stand like a rock."

—Thomas Jefferson (1762-1825)
American 3rd US President (1801-09),
Author of the Declaration of Independence

In order to have an electric current, the above two requirements can be restated as:

a) Existence of a *source of emf*, and
b) Existence of a *closed path*, between the two opposite terminals of the source of emf.

The term "electromotive force (emf)" is an extremely important term for understanding electricity, and needs to be grasped exactly before further progress in this subject.

We shall define electromotive force (emf) as, *"The difference in electric potential that exists between two dissimilar electrodes immersed in the same electrolyte, which has many ionic particles. The difference in potential causes a flow of energy, in terms of an electric current, from the higher potential to the lower."*

The method of generation of electricity, through the use of bimetallic contacts, was the first time that Man had ever produced emf. It was actually discovered by Volta (an Italian Physicist) in 1800 AD and many other methods of generating emf, such as photovoltaic, thermo-electric, electromagnetic and others, have been accomplished since then.

The existence of emf, which is the main propeller of electrons, is responsible for and the reason behind why electric current flows in the connecting wires from source to an electric load, such as a resistor in a light bulb.

"The negative electric fluid is constructed of grains, just as the beach is composed of grains of sand, or a house built of bricks. The elementary quantity of negative electricity is called electrons."

— Albert Einstein (1879-1955)
German Born American Physicist, Nobel Prize in 1921

Magnetism

The word magnetism comes from "Magnesia," an ancient city in Greece where loadstone was discovered. Loadstone, also known as “lodestone,” “leading stone,” or “Hercules stone,” is a magnetic variety of magnetite, which itself is black iron oxide (Fe_3O_4), an important iron ore and unmagnetized in the raw state. Lodestones (unlike magnetites) possess magnetic properties and attract iron or similar metals.

Magnetism, in general, is defined to be the property of certain substances (such as lodestone, etc.) that attract certain other substances (such as iron, steel, nickel, cobalt, chromium, etc.). By definition, a magnet is a body composed of a substance that exhibits magnetism, which is the ability to attract other metallic substances.

“Conscience is our magnetic compass; reason our chart.”

—Joseph Cook (1860-1947)
Australian Politician and Prime Minister

In general, there are two types of magnets: "Natural Magnets," and "Artificial Magnets."

Natural Magnets

A piece of lodestone is a natural magnet because it exhibits magnetism in its native or natural state. This property of loadstones was known to the ancients but had not been explored further until the latter part of the 16th century and the beginning of 17th century, when William Gilbert carried out the first important scientific work on this subject.

Artificial Magnets

There are materials such as iron or steel, while not naturally having the magnetic property, can be made to acquire it. These substances are called artificial magnets.

There are various ways of artificially inducing magnetism in un-magnetized iron or steel as follows:

a) Stroking or rubbing a piece of iron or steel with a strong magnet,
b) Bringing a piece of iron or steel in contact or merely near a permanent magnet (not touching) and tapping it gently a number of times. This is a process commonly referred to as induction where there is no physical contact but magnetization still takes place, and
c) Sending a current through an insulated coil of wire, which is wound around a core (the substance to be magnetized). The electric current creates a magnetic field, which causes the core material to be magnetized. This type of magnet is called an electromagnet.

The origins of magnetism, initially existing as a separate subject, later on was found to be related to electricity, in fact, its very dual universe.

There were major discoveries that led to the integration of magnetism into the subject of electricity, which led to the birth of a totally new subject called "Electro-Magnetism."

"I believe that there is a subtle magnetism in Nature, which, if we unconsciously yield to it, will direct us aright."

—Henry David Thoreau (1817-1862)
American Essayist, Poet and Philosopher

The Universal Nature of Force

In 1832, Faraday presented his vision of the physical universe, where all world forces such as electrical, magnetic, gravitational, chemical, and others existed as various forms of a fundamental force. He considered that all natural forces were essentially identical in their origin and could be converted from one form to another.

In a series of experiments, he actually showed that he could get the same chemical and magnetic effects regardless of the source of electricity. He developed the relation between electricity and magnetism and the propagation of their effects.

Faraday had a philosophical approach to his work and believed in the "unity of physical phenomena," in particular, he considered that *matter, electricity, and light were of the same origin.*

Faraday's philosophical considerations about matter, electricity, and light conform to the modern discoveries, since we know today that these are all derived from "energy," each having a progressively higher frequency with a corresponding decreasing wavelength.

His metaphysical probing into the ultimate reality of force and matter were to have a great impact on the future development of electromagnetism. Faraday saw iron filings arrange themselves in a pattern when he sprinkled them on paper held over the poles of a magnet. Faraday conceived that the *patterns were the representation of the real lines of an underlying invisible field of force existing in space.*

"Peace cannot be kept by force; it can only be achieved by understanding."

—Albert Einstein (1879-1955)
German Born American Physicist, Nobel Prize in 1921

He envisioned that a basal field of force (such as one exhibited by magnetism) accounted for the substance and operation of all nature. His vision was based on the oldest and yet continuing quest in the philosophy of science— the **Basal Unity in Nature,** i.e. *all forces or energy sources come from a common source and, therefore, are convertible into one another as shown on the next page.*

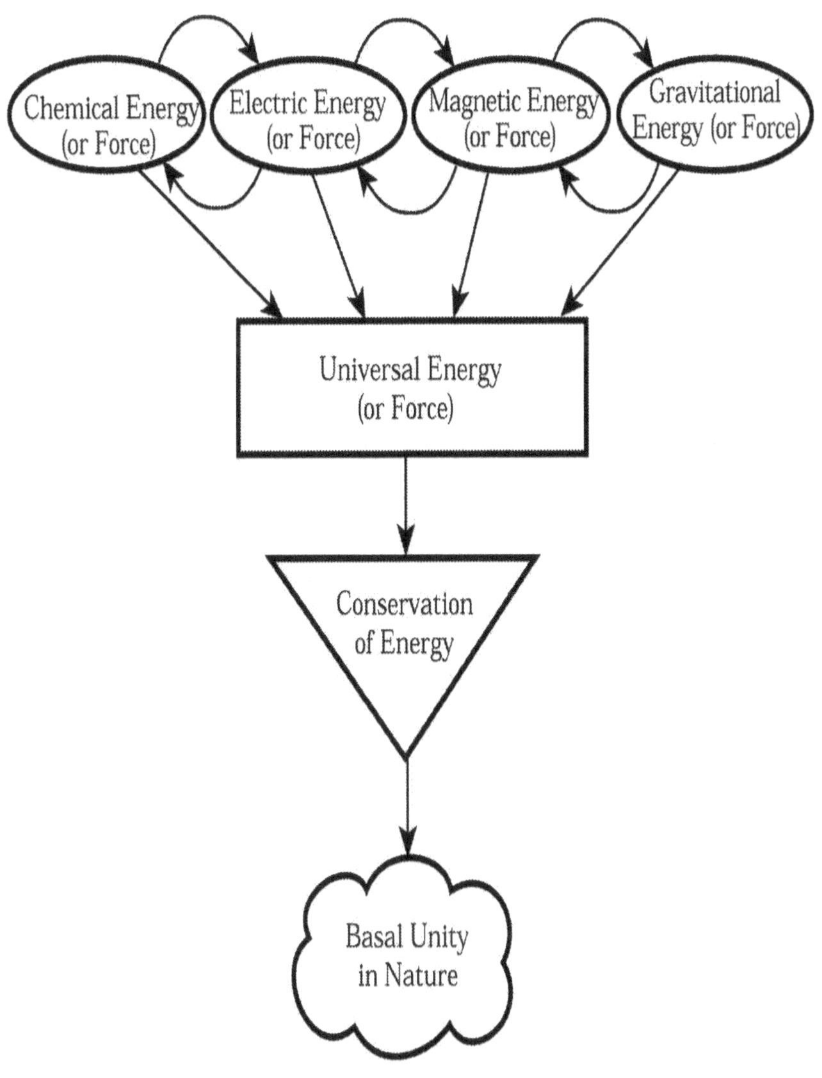

The Basal Unity In Nature.

Faraday viewed static charge and magnetism acting via a field of force occupying the whole space, which progressed continuously from force point to force point. He believed the attractions and repulsions of static electricity and magnetism were due to fields between the force centers, which caused physical deformation, deflection or strain (change in length).

Physical substance itself could be considered to be a convergence or focus of force fields. Thus, Faraday's force field theory abolished the distinction between matter and energy, since it was later discovered that matter is a condensed form of energy. Thus, we can conclude that all is derived from energy. Using this new understanding of matter and energy, Faraday rejected instantaneous action at-a-distance central force theory of Newton.

He advocated the more modern concept of force transference impelled through space as time progressed; each element impacting on a contiguous neighboring element.

His theory provided for *curved progression* of attractive or repulsive forces (rather than straight line as proposed by Newton), since they involve forces of tension and compression.

"***In questions of science, the authority of a thousand is not worth the humble reasoning of a single individual.***"

—Galileo Galilei (1564-1642)
Italian Philosopher, Astronomer and Mathematician,
who made fundamental contributions to the development of the
scientific methodology and to the sciences of motion and astronomy.

Fields

The discovery of "fields" was an important milestone principle that helped to explain the mechanism by which electricity interacted with matter. This discovery was clearly the work of a pragmatic genius, named Michael Faraday, who through observation and mental foresight was able to discover the nature of energy and how its distribution in space was an important concept in the evolution toward a complete understanding of electricity.

Faraday observed that energy, such as electricity or magnetism, do not act or apply force on matter directly. This monumental discovery was about energy acting on matter via an intermediary agent called a "field." Thus the concept of "*Field Theory*" was born.

Field-of-force theory was such an impelling and all-inclusive concept that within a few years of its publication, it led to the theoretical discovery of a major electrical principle, the propagation of waves by Maxwell.

The term *Field* has several definitions but the definition that applies here is *"An invisible entity that is distributed over a region of space and whose properties are a function of space and time. It acts as an intermediary agent in interactions between energy and matter particles."* This is shown in the Figure on the next page.

Therefore, we can now define **Field Theory** as, "*The concept that, within a space in the vicinity of a particle, there exists a field containing energy and momentum, and that this field interacts with neighboring particles and their fields.*"

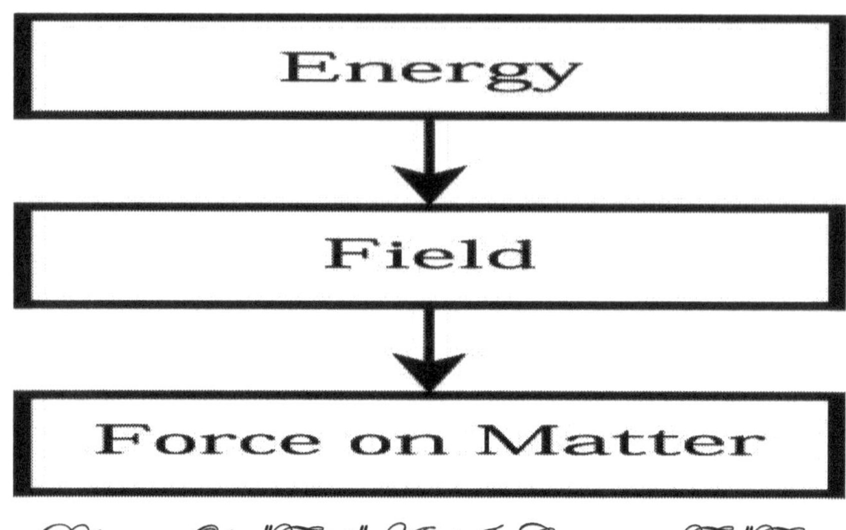

Definition Of "Field" And Its Connection To "Force

Knowing the concepts of fields, the Newtonian concepts of force were found to be deficient and had to be supplanted by the more factual concept of "**Field of Force Theory**."

Since the time of Faraday, the concept of field has expanded into many areas of human knowledge (such as gravitational phenomena, fluid flow, spirituality, etc.) and many abstruse phenomena can be explained through its utilization.

Therefore, knowing the exact definition of "Field" along with the reasons why the field-of-force theory is needed, we are well equipped to make a foray into any scientific subject and make a rapid and efficient study of it without any loss of composure.

"Energy and persistence conquer all things."

—Benjamin Franklin (1706-1790)
American Statesman, Scientist, Philosopher, Printer, Writer and Inventor

Gravity

Gravity is an important form of energy since we encounter it in every aspect of our daily existence, knowingly or unknowingly! It has made it possible for us to create a civilization in the material universe with some degree of stability, which would not lift off one day afloat in the blue yonder. We can carry on living on this planet without a concern of being hurled into the dark frontiers of outer space.

Gravity could be defined as the tendency of every mass or particle of matter to attract and be attracted by every other mass or particle. It is the mutual attraction between all masses in the universe.

"Five things constitute perfect virtue: gravity, magnanimity, earnestness, sincerity, kindness"

—Confucius (551-479 BC)

China's Most Famous Teacher, Philosopher, and Political Theorist

Toward the latter part of the 1600s AD, Sir Isaac Newton (English mathematician and philosopher) developed the laws of gravity and motion. One of his most important contributions was the "Law of Gravitation," which states that, "*Every two particles of matter in the universe attract each other with a force that acts along the line joining them, and has a magnitude proportional to the product of their masses and inversely proportional to the square of the distance between them.*"

Therefore the concept of gravity leads to the "Gravitational force," as elegantly described by Newton. However, his work was expanded to include the concept of fields, which led to the "Gravitational Field Theory."

Gravitational Field Theory is the modern theory of gravity in which gravity is treated as a "Field of Force" rather than an "Instantaneous Force Acting at a Distance," as proposed by Newton.

Gravity was a concept at the beginning of the physical universe; however, it took trillions of years to materialize and solidify as a law and thus, become a reality for us to observe, at this late stage of development of the physical universe.

Gravity has made inhabitation of planets in a body form possible; however, to some it forms an unbreakable and yet invisible chain that binds us to a planet, and with a stretch of imagination, one may see that it has made us feel like a prisoner left alone to our own devices in this lonely corner of the universe! It has made us feel like we are bound to earth eternally, without ever giving us a chance to travel to outer space freely; thus, forbidding us to navigate the rest of our galaxy let alone our immense universe.

"Love is metaphysical gravity."

— Richard Buckminster Fuller (1895-1983)
American Engineer and Architect

Heat

The concept of heat deals with the amount of energy contained in an object, a fluid, a gas, or a plasma. The temperature of an object is the degree of its hotness or coldness and, therefore, is an index of the amount of energy (in the form of heat) contained in that object.

To be exact, heat is considered to be, "*A form of energy whose effect is produced by the accelerated vibration of molecules and into which mechanical energy may be converted.*" Theoretically speaking, at zero degrees Kelvin or -273.16°C molecular vibration would stop and there would be no heat!

Heat transfer is, "*Energy in transit due to a temperature difference between the source, which the energy is coming from, and a sink toward which the energy is going.*"

Heat transfers through several mechanisms, such as:

a) *Conductive Heat Transfer*: A process by which heat diffuses through a solid or a nonmoving fluid. An example would be the heat transfer in a heat sink attached to a transistor.

b) *Convective Heat Transfer*: A process by which the mass movement of parts of a fluid within the fluid because of differences in the temperature of the parts. This is where a **moving** fluid transfers heat to or from a location. An example would be the heat transfer process caused by a fluid moving in a closed circuit, such as in a car radiator.

c) *Radiative Heat Transfer*: A process by which long-wavelength electromagnetic waves transport heat from a solid, fluid or gas, as a result of their temperature (also called thermal radiation). An example would be the heat transfer provided by intense radiation, as in a heat lamp or the sun.

An important branch of physics is "**Thermodynamics,**" *that deals with the transformation of heat into other forms of energy (particularly mechanical energy) and studies the laws governing such conversions of energy.*

Stepping back and examining our existence, we see that we are dealing with a thermodynamic machine (our bodies) which converts food into motion and other forms of energy and thus, provides us with a certain level of survival and stability on a physical plane.

However, as fascinating as the body as an energy-converting machine may be in providing us with an orientation point in the physical universe, nevertheless to some, it may seem that it has us at its beck and call on a continuous basis with its constant daily demands for food, clothing, shelter, etc., thus, precluding us to achieve a higher level of mental awareness or to soar to new heights of existence.

"The thing to do is to supply light and not heat."

—Woodrow T. Wilson (1856-1924)
American 28th President of the United States

Waves

"Just as the wave cannot exist for itself, but is ever a part of the heaving surface of the ocean, so must I never live my life for itself, but always in the experience which is going on around me."

—Albert Schweitzer (1875-1965)

German Medical Missionary, Theologian, Musician and Philosopher
1952 Nobel Peace Prize

We define waves as, "A disturbance that propagates from one point in a medium to other points without giving the medium as a whole, any permanent displacement."

Considering the concept of waves, we realize that they are part and parcel of this universe. Some waves such as light waves are actually necessary for sustaining life forms and organisms since the beginning of life on earth.

We are dealing with the physical universe, a universe existing due to and founded upon motion primarily, and quite visibly, energy secondarily. This is true because every piece of it is in constant motion microscopically if not macroscopically.

"We can speak without voice to the trees and the clouds and the waves of the sea. Without words they respond through the rustling of leaves and the moving of clouds and the murmuring of the sea."

—Paul Tillich (1886-1965)

German Born American Theologian and Philosopher

By observation, we note that waves are a form of kinetic energy, which deals with motion of objects or particles. According to the modern theory of waves, a wave can be associated with any moving object. The corresponding wavelength is inversely proportional to the speed of the object. This means that a faster object has an associated wave with a smaller wavelength, and as a result, a higher frequency of operation.

There are two main types of waves:

a) Mechanical waves, and
b) Electromagnetic (or EM) waves.

By ***mechanical wave*** we mean a wave that is caused by the vibration of matter particles as it propagates in a medium. Examples of this type of waves are sound waves, earthquake waves, etc.

If the motions of matter particles conveying the wave are perpendicular to the direction of propagation of the wave itself, we then have a *transverse wave*. For example, when a string under tension is set oscillating back and forth at one end, a transverse wave travels down the string to the other end; the disturbance moves along the string but the string particles move perpendicular to the direction of propagation of the disturbance.

If the motion of the particles conveying a mechanical wave is back and forth along the direction of propagation, we then have a *longitudinal wave*. For example, when a vertical spring under tension is set oscillating up and down at one end, a longitudinal wave travels along the spring; the coils vibrate back and forth in the direction in which the disturbance travels along the spring (much like the shock absorbers in a car oscillating while going over a bump). Sound waves are also longitudinal waves.

"“The winds and the waves are always on the side of the ablest navigators."

—Edward Gibbon (1737-1794)
English Historian

Electromagnetic (or EM) wave is a disturbance which consists of oscillating electric and magnetic fields, propagating at the speed of light, which is 186,000 miles per second. The two fields are at right angles to each other and lie in the transverse plane, which is a plane perpendicular to the direction of propagation. For this reason EM waves are also classified as "*transverse waves*."

EM waves (such as light) have existed and have been utilized by life forms but their nature was not analyzed or understood until very recently when James Clerk Maxwell (1831-1879) formulated his EM theory and presented it as a scientific and solid foundation for understanding waves circa 1873.

Maxwell's work clearly showed that the source of EM waves, particularly light waves, is electricity, quite a startling conclusion at the time. Over two decades later in 1894, Hertz's the discovery of EM waves experimentally proved Maxwell's mathematical prediction of their existence.

Maxwell's work demonstrated that light waves were actually electromagnetic waves caused by electric sources, a concept totally new and revolutionary, which collided with orthodox optics and its established laws. Maxwell's discovery, though coolly received at the time by scientists, laid the cornerstone for all future discoveries in the field of radio telecommunication, electro-optics, electro-magnetism, radar, and many other fields yet to be discovered. Applications of EM waves have changed the course of our civilization and set it on a totally new track.

Thus, we are dealing with an old and yet a very familiar topic but with a totally new understanding of what waves are and how they behave under different conditions and in different media.

"Life is a wave, which in no two consecutive moments of its existence is composed of the same particles."

— John Tyndall (1820-1893)
Irish Scientist

Radio Waves and Microwaves

Radio waves (also called Radio Frequency waves, or RF waves for short) are electromagnetic (EM) waves propagating in the frequency range of about 10 kHz (10^4 Hz) to 1 GHz (10^9 Hz). This is the mechanism by which energy transfers through space by means of Electromagnetic radiation at radio frequencies.

Microwaves are EM waves that exist at higher frequencies ranging from 1GHz to 300 GHz having wavelengths ranging from 30 cm (at 1 GHz) to 1 millimeter (at 300 GHz).

It should be noted that there are no sharp boundaries distinguishing microwaves from RF at the lower end and from infrared at the upper end of the frequency spectrum.

Properties of RF/Microwaves

An important property of signals at RF, and particularly at higher microwave frequencies is their great capacity in carrying information. This is due to the existence of large bandwidths that are available at these high frequencies. For example, a 10% bandwidth at 60 MHz carrier signal is 6 MHz, which is approximately one TV channel of information; on the other hand, 10% of a microwave carrier signal at 60 GHz is 6 GHz which is equivalent to 1000 TV channels.

Another property of microwaves is that they travel by line of sight, very much like traveling of light rays as described in the field of geometrical optics. Furthermore, unlike the lower frequency signals, the microwave signals are not bent by the ionosphere. Thus, use of line-of-sight communication towers or links on the ground and orbiting satellites around the Globe are a necessity for local or global communications.

A very important civilian, as well as a military instrument is radar. The concept of radar is based upon radar cross-section which is the effective reflection area of the target. Target's visibility greatly depends on the target's electrical size which is a function of the incident signal's wavelength. Microwave frequencies are the ideal signal band for radar applications. Of course, another important advantage for the use of microwaves in radars is the availability of higher antenna gains as the frequency is increased for a given physical antenna size. This is because the antenna gain is proportional to the electrical size of the antenna, which becomes larger as frequency is increased in the microwave band. The key factor in all this is that microwave wavelengths are comparable to the physical size of the transmitting antenna as well as the target.

There is a fourth and yet a very important property of microwaves and that is the molecular, atomic, and nuclear resonance of conductive materials and substances when exposed to microwave fields. This property creates a wide variety of applications. For example, since almost all biological units are composed of water predominantly, and as we know water is a good conductor, thus, microwave gains tremendous importance in the field of detection, diagnostics, and treatment of biological problems or investigations as in medicine (e.g. diathermy, scanning, etc.). There are other areas that this basic property would create a variety of applications such as remote sensing, heating (e.g. industrial purification, cooking, etc.), and many others.

Reasons for Using RF/Microwaves

Over the past several decades, there has been a growing trend toward the use of RF/Microwaves in system applications. The reasons are many, amongst which the following are prominent:

a) Wider bandwidths due to higher frequency
b) Smaller component size leading to smaller systems
c) More available and less crowded frequency spectrum
d) Better resolution for radars due to smaller wavelengths
e) Lower interference due to a lower signal crowding
f) Higher speed of operation
g) Higher antenna gain possible in a smaller space

On the other hand, there are some disadvantages in using RF/Microwaves such as: use of more expensive components; availability of lower power levels; existence of higher signal losses, and use of high-speed semiconductors (such as GaAs or InP) along with their corresponding less-mature technology, relative to the traditional Silicon technology which is quite mature and less expensive at this time.

In many RF/Microwave applications the need for a system operating at these frequencies with all the above advantages, is so great that it outweighs these disadvantages and spurs the engineer forward into a high-frequency design.

RF/Microwave Applications

The major applications of RF/Microwave signals can be categorized as follows:

I. Communication: This application includes satellite, space, long-distance telephone, marine, cellular telephone, data, mobile phone, aircraft, vehicle, personal and Wireless Local Area Network (WLAN), and so on. One important sub-category of this application is TV and Radio broadcast

II. Radar: This application includes air defense, aircraft/ship guidance, smart weapons, police, weather, collision avoidance, imaging, etc.

III. Navigation: This application is used for orientation and guidance of aircraft, ships, and land vehicles. Particular applications in this area are:

a) Microwave Landing System (MLS), which is used to guide aircraft to land properly in airports.

b) Global Positioning Systems (GPS), which is used to find one's exact coordinates on the Globe.

IV. Remote Sensing: In this application, many satellites are used to monitor the Globe constantly for weather conditions, meteorology, ozone, soil moisture, agriculture, crop protection from frost, forests, snow thickness, icebergs, and other factors such as natural resources monitoring and exploration, etc.

V. Domestic and industrial applications: This application includes microwave ovens, microwave clothes dryers, fluid heating, moisture sensors, tank gauges, automatic door openers, automatic toll collection, highway traffic monitoring and control, chip defect detection, flow meters, power transmission in space, food preservation, pest control, etc.

VI. Medical applications: This application includes cautery, selective heating, heart stimulation, hemorrhage control, sterilization, imaging, etc.

VII. Surveillance: This application includes security systems, intruder detection, electronic warfare (EW) receivers to monitor signal traffic, etc.

VIII. Astronomy and space exploration: In this application, gigantic dish antennas are used to monitor, collect, and record incoming microwave signals from outer space, providing vital information about other planets, star, meteors, etc., in this or other galaxies.

IX. Wireless applications: Short-distance communication inside as well as between buildings in a local area network (LAN) arrangement can be accomplished using RF and Microwaves. Connecting buildings via cables (e.g. coax or fiber optic) creates serious problems in congested metropolitan areas, since the cable has to be run underground from upper floors of one building to upper floors of another. However, this problem can be greatly alleviated using RF and microwave transmitter/receiver systems which are mounted on rooftops or in office windows. Inside buildings, RF and Microwaves can be used effectively to create a wireless LAN in order to connect telephones, computers, and various LANs to each other. Using wireless LANs has a major advantage in office re-arrangement where phones, computers and partitions are easily moved with no change in wiring in the wall outlets. This creates enormous flexibility and cost saving features for any business entity.

"I cannot conceive curved lines of force without the conditions of a physical existence (of energy) in that intermediate space."

—Michael Faraday (1791-1867)
English Physicist, Whose Genius Mind Contributed
Greatly to the Understanding of Electromagnetism

Light

"Someday perhaps the inner light will shine forth from us, and then we'll need no other light."

—Johann W. von Goethe (1749-1832)
German Poet and Novelist

When we say "light" we actually mean "visible light," which is, "*An electromagnetic wave (or radiation) traveling at the speed of 186,000 miles per second in a vacuum and with a wavelength between 400 and 770 nanometers (nm) perceptible by the human eye.*"

Visible light consists of six main colors (wavelengths), which can be listed in the sequence of decreasing wavelength as:
a) Red,
b) Orange,
c) Yellow,
d) Green,
e) Blue, and
f) Violet.

Although ultraviolet and infra-red radiations may barely excite some of the photocells in the eye, they are usually not considered light.

However, we could "generalize" the term "light" to include radiations in the infrared as well as ultraviolet bands ranging from 100 nm (ultraviolet) to 10,000 nm (Infrared) in wavelengths.

The human eye thus eliminates a large amount of information from our daily experience that we could have otherwise received. Living, as a human being, is a limited information type activity, which creates many blind sides to existence.

Furthermore, the human eye has a spatial filter, which limits the angle of information to less than 180°. Thus, not being omnivision (seeing 360°) causes another series of limitations on our functionality and productivity in our lives and further complicates our decision power and judgment in many situations.

Therefore, *being human is being limited*, which creates many trials and tribulations in our daily existence due to an enormous amount of optical information that a body filters out. Thus we have to make up for this serious deficiency on a constant basis either through visualization or the use of advanced equipment.

"***You must enshrine in your hearts the spiritual urge towards light and love, Wisdom and Bliss!***"

—Sri Sathya Sai Baba (1929-)
Indian Spiritual leader

Sound

Sound is, "*A mechanical wave requiring a physical media for its propagation unlike EM waves, which do not need any physical entity for their existence*."

Sound waves *are* a result of the alteration of properties of an elastic medium, such as pressure, density, etc. Sound is being a mechanical wave (a much coarser wave than light) and is dependent upon matter for its existence. It has a much lower speed of propagation compared to light (1129 ft/s or 2300 miles/ hour).

"Kindness is a language which the deaf can hear and the blind can see."

— Mark Twain (1835-1910)
American Humorist and Writer

Audible sound is a vibration in air, water, etc., that stimulates the auditory nerves and produce the sensation of hearing. It has a sonic frequency range of about 15 to 20,000 Hertz (Cycles/s), which is also called the audio range. Sound waves having vibrating frequencies above the audio range are termed "ultrasonic," while those with frequencies below are called "infrasonic."

Noting that the audio range is a very small portion of the possible sound frequencies, we realize that by using a body we, as beings, are using a sound filter, which eliminates a large amount of information that we could have received otherwise. This truly shows that being human is being limited and thus, the old adage, "It is human to err," becomes the ruling principle.

"Get someone else to blow your horn and the sound will carry twice as far."

—Will Rogers (1879-1935)
American Actor-Entertainer, Philosopher and Journalist

Chemical Energy

Another form of energy on a microscopic level is chemical energy. By chemical energy, we mean the energy residing in the chemical bonds between the atoms of a compound, which is a form of potential energy (see the Figure shown below).

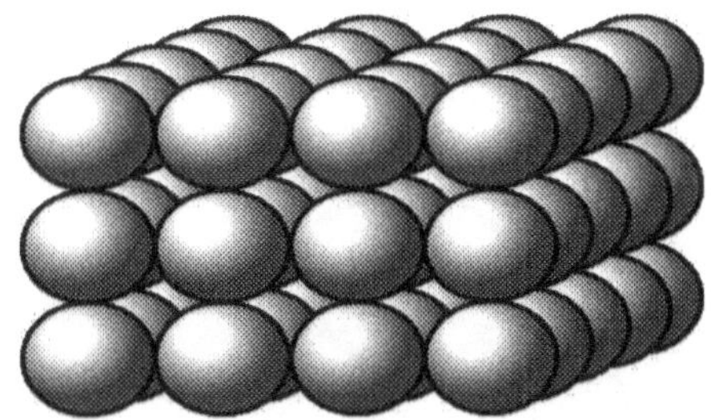

Chemical Energy

There are three main types of atomic bonding: a) Ionic bonding (abduction of electrons causes ions to form and attract each other), b) Metallic bonding (metal ions in a sea of electrons), and c) Covalent bonding (atoms sharing electrons with each other).

When two compounds interact with each other, the atoms are rearranged in the reacting compounds producing new compounds and heat. The energy of the whole system is conserved according to the principle of conservation of energy, thus, the heat produced is a form of released energy (resident as potential energy) in the original compounds that is being converted to kinetic energy.

"Real wealth is ideas plus energy."

— Richard Buckminster Fuller (1895-1983)
American Engineer and Architect

Nuclear Energy

Considering energy on a microscopic level, we have to consider nuclear energy as an important form of energy. Nuclear energy (also known as atomic energy) is, "*That form of energy resident in an atom that is released through either nuclear fission or fusion.*"

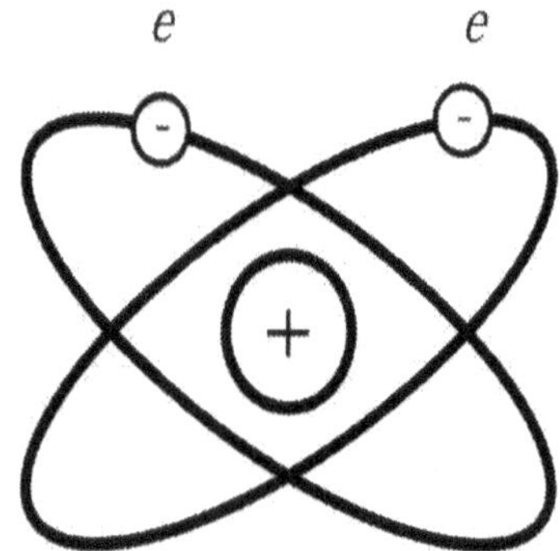

The Potential Energy In The Atom.

By **nuclear fission,** we mean the division or splitting of an atomic nucleus (usually heavier nuclei such as Uranium, Plutonium, etc.) into two fragments of approximately equal mass accompanied by the conversion of part of the mass into energy. This is the process of converting resident potential energy in the atomic structure into kinetic energy such as heat. This is the principle that is being used in the construction of the atom bomb, which has opened a dark chapter in human history.

A more peaceful application of this principle is used in the production of electricity in a nuclear reactor. A nuclear reactor is a device containing fissionable material (called the fuel element) in sufficient quantity and so arranged as to be capable of maintaining a controlled, self-sustaining nuclear fission chain reaction.

By **nuclear fusion,** we mean the fusion or combination of two light nuclei (such as Deuterium, Tritium, etc.) to form a nucleus of heavier mass (such as Helium) with a resultant loss in the combined mass, which is converted into energy. This is the principle used in the construction of Hydrogen bomb with its many destructive results.

Exploration of energy at such small levels with such an enormity of energy release, has opened many doors for mankind for potential applications in our lives but its destructive side of applications by the anti-social beings has brought forth once again a major dilemma posed by our contemporary philosophers and that is "should man have access to such enormous energy sources for destruction without ever being improved ethically?"

In other words, through the use of nuclear energy, we have unleashed an enormous source of energy into our midst, which could be used as an unlimited weapon, without ever conceiving a defense against it. The ultimate defense against such an unlimited weapon is not another machine or another materially conceived contraption but an honest use of the principles of ethics first and foremost amongst our political leaders and heads of states, who are going to make the ultimate decisions, and secondarily amongst our scientist and engineers who are irresponsibly inventing these unlimited weapons.

"We all need to look into the dark side of our nature - that's where the energy is, the passion. People are afraid of that because it holds pieces of us we're busy denying."

—Sue Grafton (1940-)
American Writer

PART VII

Understanding Matter

Matter

Matter and energy were considered separate entities for a long time until recent discoveries and advances, which made them equivalent. Thus, based on this understanding we need to define matter at this point as, "*The result of bringing energy particles into close proximity where they occupy a very small volume.*"

For example, we need to bring electrons into close proximity with a nucleus (protons and neutrons) to create a matter particle called an atom. Conversely, matter becomes energy if dispersed or decompressed. For example, in a nuclear fission process, matter is converted into raw energy, which can be harnessed to create electricity. This concept is shown below.

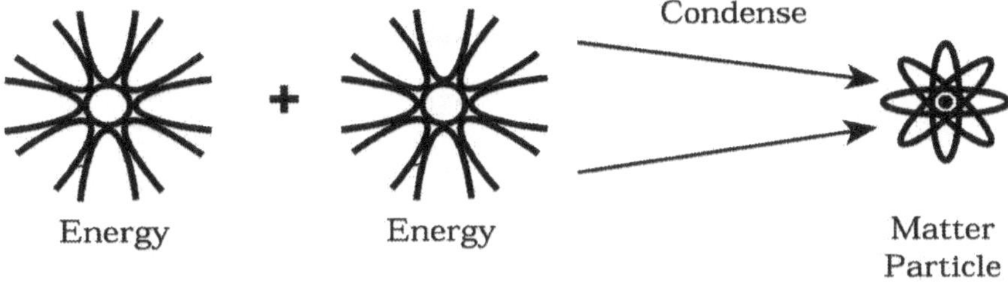

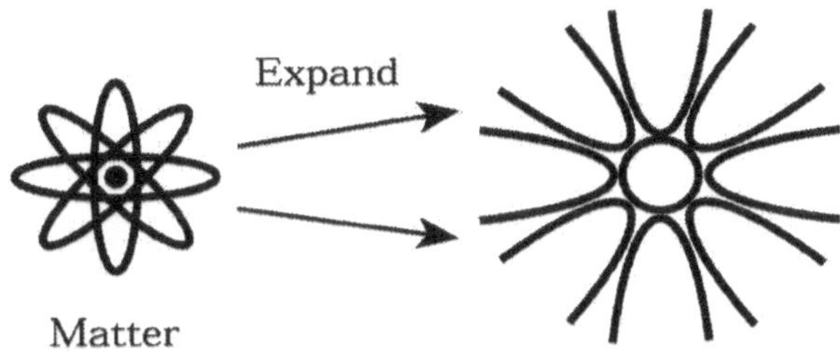

Relationship Of Matter And Energy.
[Ref. 8, pp.10-13]

Mass is a fundamental property of a body, which determines the acceleration of the body when a given force is applied. Mass, a measure of the quantity of matter, is only one aspect that describes the matter quantitatively and is usually considered to be a constant, although it varies with the velocity of the body. Mass and energy are interchangeable by the Einstein's relationship: $E=mc^2$, which essentially states that energy (E) is basically equivalent to mass (m), by a proportionality factor of c^2.

Conversion of mass to energy is not the same as, "combustion" as is done in most modern engines. Since through combustion of the fuel with oxygen, the energy stored in the chemical bonds are released and new chemicals are formed with total mass at the end of the process remaining almost the same.

Actual conversion of the molecules of the fuel into energy has no byproducts with the final mass being zero. For example, converting one kilogram of mass produces approximately $9x10^{16}$ joules of energy which can light up Los Angeles for more than ten years! This fact shows our current inefficient methods of creating energy, which gives us an energy crisis every few years and a polluted planet.

We can observe that any piece of matter, on a classical physics level of observation, could be considered to be a standing wave in its last stage of suspension in space. Standing waves have a quite long duration and can exist long after the flow itself has ceased to exist. This is analogous to memory recordings of an incident in one's memory banks where one can recall and find out what happened long after the force of the impact has disappeared!

"If a theory is complicated, it's wrong."

— Richard Feynman (1918-1988)
American theoretical physicist

Particle

It is understood that when we refer to a particle, we mean a piece of matter so small as to be considered without size, though having inertia and the force of attraction.

More precisely, we can define a particle as, "*Any infinitesimal subdivision of matter, which is ranging in diameter from a fraction of an Angstrom (such as an electron, an atom or a molecule, etc.) to a few millimeters (such as a raindrop, etc.).* "This concept is shown below.

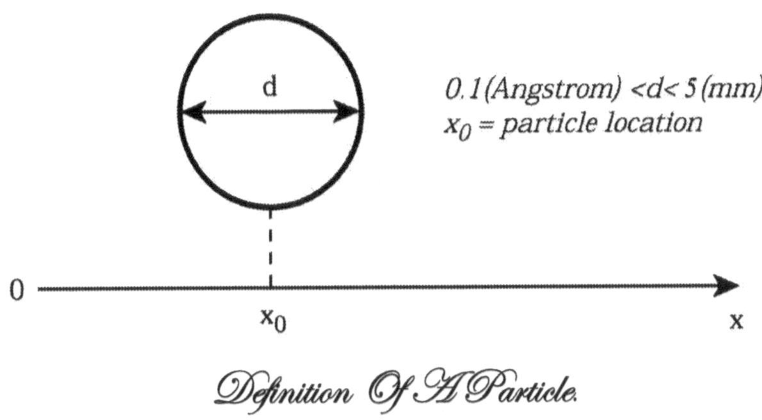

Definition Of A Particle

Our universe is made up of an almost infinite number of particles, which when combined into building blocks, can create any and all four forms of matter: solids, liquids, gases and plasmas. These four forms of matter create all of the expansive masses of objects, cities, countries, planets, galaxies, and the entire physical universe.

"The particle and the planet are subject to the same laws and what is learned of one will be known of the other."

—James Smithson (1765-1829)
English Chemist, founder of the Smithsonian Institute in Washington D.C.

Sub-Atomic Matter

Sub-Atomic Matter (or sub-Microscopic Matter) is actually the condensation of energy particles to create light energy particles that will be utilized in the creation of the basic unit of stable matter commonly referred to as an atom. The subatomic particles comprising an atom are in such a state of uncertainty, agitation, and constant confusion on the microscopic or quantum level, that it would be difficult to track down any of them accurately and thus, rendering its nature completely occluded, uncertain, and undetermined (due to their unpredictable motion).

Examples of subatomic particles are electrons, quarks, neutrons, protons, etc. Amongst these particles, "quark" seems to be the most fundamental subatomic particle.

In other words, when we enter the realm below the level of the constituent particles, which make up the subatomic particles, we are apt to run into a whole level of uncertainty, unreality, and mystery from which a particle physicist has yet to escape! This fact alone makes anything made from any of the subatomic or elementary particles (which practically includes every atom, molecule, object, planet, etc.) a total uncertainty at its core and thus, gives us a mysteriously indeterminate universe, to say the least!

"Enthusiasm is the divine particle in our composition: with it we are great, generous, and true; without it, we are little, false, and mean."

—Letitia Elizabeth Landon (1802-1838)
English Poet and Writer

Quark

We can define quark as, "*A basic particle having a fraction, 1/3 or 2/3, of the charge of an electron from which many of the elementary particles such as electrons, protons, or neutrons may be built up.*"

The quarks have never existed in isolation but have remained confined within protons, neutrons and other particles. In other words, free quarks are purely theoretical particles and by experimental evidence, they cannot exist *freely* as independent particles. Even if they exist momentarily, they may prove to be highly unstable and will combine to form more stable particles. Of course, this may be contrary to the current belief of the particle physicist for whom quarks may be very real particles.

Even though, the existence of a quark as a particle has been proven experimentally by indirect methods at this point in time, the possibility of existence of particles smaller than quarks still exist. This is because of the non-absoluteness axiom, which states that there are no absolutes in the physical universe. Therefore, to call a quark the smallest particle in the universe brings forth an absolute, which is an impossibility!

The more realistic picture would be an understanding of the fact that any series of particles smaller than quarks, whose size approach zero (but never achieving zero) is possible, awaiting a future discovery someday.

Needless to say, particles at these microscopic levels look more like compressed or condensed waves than points of solid matter.

"An approximate answer to the right problem is worth a good deal more than an exact answer to an approximate problem."

—John Wilder Tukey (1915-2000)
American Mathematician and Scientist

Electron

"Electron" is an important particle, which needs to be grasped exactly, with some possible clarification about its role in the existence of many electrical phenomena that we see about us.

Electron is defined as, *"A stable elementary particle of matter, which carries a negative electric charge of one electronic unit equal to* $q = -1.602 \times 10^{-19}$ *C and has a mass of about* 9.11×10^{-31} *kg and a spin of ½."*

We need to clear up a general misconception about electrons, which is worthy to note at this point. An electron is a "*charge carrier*" with an inherent negative charge but itself is not the "*charge*" but a particle with a mass. As such, it has its own mechanical type of energy (such as kinetic and/or potential), which is in addition to its charge. In simple terms, we can observe that charge is electric energy manifesting itself in terms of a built-in electric force field, and an electron is not the charge but a charge carrier.

We assign a negative value to the type of charge that is carried by an electron and a positive value to the charge of a proton. However, in actual practice, since the only mobile charges are electrons, a **surplus of electrons** at a location means a **net negative charge** at that point in space. This is because they outnumber the protons. On the other hand, the **absence of electrons** indicates the existence of a **net positive charge**, since the remaining electrons become outnumbered by the proton charges.

"What is a soul? It's like electricity - we don't really know what it is, but it's a force that can light a room."

—Ray Charles (1930-2004)
American Pianist and Singer

Proton

The opposite of an electron is a proton, which is a particle located at the center of an atom and around which electrons revolve as they spin around themselves. It is the pivotal point around which the whole world of an atom exists.

It is defined as, "*An elementary particle of matter that is the positively charged constituent of ordinary matter and together with the neutron, is a building block of all atomic nuclei; its rest mass is* $1.67x10^{-27}$ *Kg (which is 1833 times bigger than that of an electron) and a spin of ½.*"

"The rapidity of particle flow alone determines power."

—L. Ron Hubbard (1911-1986)
American Philosopher, Writer, Educator, and Humanitarian

A proton is the *central stable datum* around which the enormous confusion, caused by the motion of electrons, is supported. Its positive charge as well as its mass is many times that of the electrons orbiting it; thus providing the much needed attractive force (electrical and gravitational) to bind the electrons to the nucleus. This creates a tightly-packed package and the smallest building block (i.e., an atom), out of which the whole panoramic universe is built.

"The discovery of nuclear reactions need not bring about the destruction of mankind any more than the discovery of matches."

—Albert Einstein (1879-1955)
German Born American Physicist, Nobel Prize in 1921

Neutron

***A Neutron** is, "One of the uncharged stable elementary particles of an atom having the same mass as a proton. A free neutron decomposes into a proton, an electron, and a neutrino. A neutrino is a neutral, uncharged particle but is an unstable particle since it has a mass that approaches zero very rapidly (a half-life of about 13 minutes)."*

All matter particles are composed of atoms, which themselves are composed of electrons (carrying a negative charge), protons (carrying a positive charge) and neutrons. A positive charge is obtained by rubbing a glass object with silk, while a negative charge may be obtained by rubbing a piece of resin with wool.

At first glance it appears that the neutron as a basic particle is without charge, but this is only an apparency.

Through a closer examination of the neutron and its definition, the facts surface rather quickly and our statement of, "*All stable matter is composed of positive and negative charged particles,"* is verified.

"Ours is a world of nuclear giants and ethical infants. We know more about war that we know about peace, more about killing that we know about living."

—Omar Bradley (1893-1981)
American General, commanded US ground forces
in Normandy invasion in World War II

Photon

In 1887, Hertz almost discovered photons by discovering the "photoelectric effect," which is also known as "photoelectricity." Photoelectric effect consists of throwing off of electrons by a metallic surface when exposed to a monochromatic ray of light composed of photons. If the photons are of sufficient energy, a charged plate can actually be completely discharged by focusing a beam of light on it.

"For light I go directly to the Source of light, not to any of the reflections."

—Peace Pilgrim (1908-1981)
American Spiritual Leader, and Peace Prophet

However, it was not until 1920s when Max Planck observed that radiation from a heated black body at a fixed temperature (also known as a black body radiation) is emitted in discrete units of energy called quanta (plural of quantum). Each quantum of energy (E) for a monochromatic light is proportional to its frequency (f) through a proportionality constant called the Planck's constant ($h=6.63x10^{-34}$ Js) .

Soon after Planck developed his hypothesis of the quantum nature of radiation from a black body, Albert Einstein performed an experiment involving the photoelectric effect, and demonstrated the quantized or discrete nature of light.

"The more success the quantum theory has, the sillier it looks."

—Albert Einstein (1879-1955)
German Born American Physicist, Nobel Prize in 1921

Einstein's experiment clearly verified the correctness of Planck's hypothesis and brought forth the quantum theory of light. The quantized units of light can be considered to be localized packets of energy called “photons.” This aspect of light, is quite contradictory to the classical physics view, which regards light in terms of a wave.

More precisely we can define a Photon (also known as a light quantum) as, “A particle, with a negligible mass (zero relative to an electron), which carries energy, linear momentum, and angular momentum. It is the quantum of energy of an electromagnetic field.

“The difficulties you meet will resolve themselves as you advance. Proceed, and light will dawn, and shine with increasing clearness on your path.”

—Jim Rohn (1930-)
American Motivational Speaker, Philosopher and Author

Atom

An atom is made up of many internally polarized positive and negative charged sub-atomic particles (electrons, protons, etc.). The nuclear forces between particles are of gravitational as well as electric and magnetic type. The electric forces are either a) An attracting force on unlike charged particles, or b) A repelling force on like-charged particles.

Due to the inherent attractive forces involved between the positively and negatively charged particles, they cannot be allowed to collapse on each other, otherwise the space will also collapse. Thus, negative charged particles are separated by permanent motion around a central positive charge (called the nucleus) at certain speeds and at specific radial distances from it (see the Figure shown below).

The centrifugal forces caused by this motion is counteracted by the attractive forces (both electrical and gravitational), causing the electrons to be at fixed and exact orbits from the nucleus.

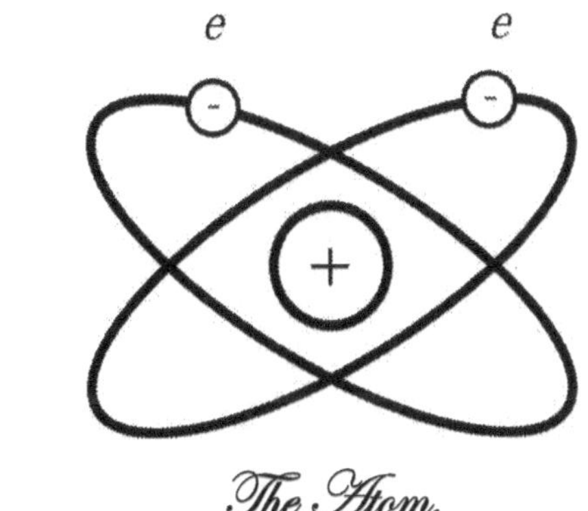

The Atom.

Different atoms can exist by taking on different number of electrons and protons and neutrons. Thus, the whole gamut of basic elements as partially outlined in the periodic table can be brought into existence. The periodic table, consisting of 103 basic and known elements on earth, is only a partial listing of all the possible elements, since the solar system (let alone earth) is an extremely small member of the physical universe.

So all of the undiscovered elements on other planets, stars, interstellar gases, and dusts form a formidable list of basic elements (creating an Expanded Periodic Table), which exist in the physical universe. The Expanded Periodic Table will include all existing basic elements in the entire physical universe. A terribly daunting task left to future generations of mankind to discover, when the entire physical universe (spanning billions of light years) is completely mapped out.

The atoms have the property of combining with the same or other atoms to form groups of atoms (such as molecules, etc.) to bring forth innumerable combinations of atoms and a host of materials found either in nature or generated synthetically. The atoms group together through a process called bonding, thus, creating three types of bonds (i.e., ionic bonding, metallic bonding, and covalent bonding) to form molecules, crystal lattices, polycrystalline materials, amorphous materials, and many other compound materials.

Just as atoms are the building blocks of inanimate matter, cells are the building blocks of animate matter. A cell is composed of organized matter plus "life force," which is an external energy form and is actually immensely different than its physical universe energy counterpart.

Thus the microscopic atoms, with their core nucleus and the orbiting electrons (internally), as well as their interaction laws with other atoms (externally), dictate the make up of our macroscopic universe; in short, *"The microscopic atoms rule the macroscopic galaxies."*

"That the universe was formed by a fortuitous concourse of atoms, I will no more believe than that the accidental jumbling of the alphabet would fall into a most ingenious treatise of philosophy."

—Jonathan Swift (1667-1745)
Irish Author and Satirist

Forms of Matter

Matter is found normally in one of four possible forms: solid, fluid, gas and plasma. Their existence in one of the four forms is primarily based upon temperature and secondarily on the density, atomic composition, etc.

These four forms of matter establish the building blocks for the subsequent creation of objects and large masses where the gigantic forces (whether gravitational, electric, magnetic, or others) of these masses, which is a collection of many sub-forces, cause them to be in constant motion relative to one another. Matter itself is made up of fast moving energy particles; therefore, every piece of matter would be perforce in constant motion microscopically, if not macroscopically.

By combining the atoms in various quantities and combinations, a great many number of complex materials with different density, texture, quality, color and shape can be created or synthesized. With the help of the expanded periodic table we can surely visualize that the possible combinations of atoms forming into complex materials will be greatly increased.

The important point to remember from this discussion is that, since the basic constituent particles of any elemental material (i.e., atoms) are made up of a series of fast, and randomly moving sub-particles with uncertain and indeterminate natures, then all of the derived byproduct materials (e.g. plastic, wood, paper, etc.) are also changing constantly internally and thus, exist on a very shaky and uncertain foundation.

The only certainty that a piece of matter (such as a metal, a dielectric, etc.), a body (such as an object, etc.), or an application mass (such as a car, a component, etc.) offers is based on an apparency of observation, which itself is fleeting and transient.

If there is any certainty on a macroscopic level of observation, it is because the viewpoint postulates it into existence and is not truly based upon the actuality or material universe facts of the situation.

Moreover, due to the constant interaction of heavy matter particles (e.g. atoms, etc.) with lighter energy particles (e.g. photons, etc.), a series of sequential actions or changes do take place. These actions or changes can be kept tack of by plotting them on a linear scale, which we call a *time stream*. Thus the concept of time is born.

"The only real valuable thing is intuition."

—Albert Einstein (1879-1955)
German Born American Physicist, Nobel Prize in 1921

Superconductors

It was discovered in the 1800s that the resistance of a material is influenced by the temperature. This recognition was the first step on a road that culminated in the achievement of zero resistance, nearly a century later. Thus, the concept of superconductivity was born.

In 1908, A Dutch physicist, by the name of H. Kamerlingh Onnes (1850-1926) in Leyden, Holland, through liquefying Helium was able to achieve a temperature of 4.2 degrees Kelvin. This was a major step in achieving temperatures close to zero degrees Kelvin, because Onnes knew from the earlier observations made by Ohm that if such low temperatures could be achieved, he could reduce or possibly, completely vanish the resistance of a material.

So, Onnes began his experimental work in 1911 and after several trials and errors on several metals, he was able to show that the resistance for Mercury would sharply drop to zero at a temperature of 4.2 K. This was a very fortunate situation, since the means to get to lower temperatures were not available at the time. Later on, Onnes found out that Lead fell to zero resistance at 7.2 K and Tin at 3.2 K. He termed this new phenomenon "*Superconductivity.*"

Other metals entered the state of zero resistance at a temperature called "critical temperature". With this discovery, Onnes realized that superconductors, having a zero resistance, could carry large amounts of current, and thus, were accompanied by large magnetic fields without the usual heating effects that accompanied non-superconductors.

"It may be that you are not yourself luminous, but you are a conductor of light. Some people without possessing genius have a remarkable power of stimulating it."

—Arthur Conan Doyle, Sr. (1859-1930)
Scottish Writer, Creator of the Detective Series, Sherlock Holmes

The possibilities of applications of superconductors through improving their performance did not occur until almost fifty years later. A breakthrough came after World War II with the discovery that alloys of Niobium, Vanadium and other transition elements could carry currents of thousands of amperes and very large magnetic fields. This breakthrough opened the door to research for other superconducting alloys that operate at higher temperatures such as 25 degrees Kelvin and higher.

One possible application of superconductors would be in the area of generation of high-intensity magnetic fields. The present day technology is based on copper-wound wires and iron-core magnetic solenoids, with the magnetic flux density saturating at 2 Teslas, which is the upper limit of conventional magnetic fields. With the manufacturing of wires made of transition-metal alloys and use of cryogenics, superconducting magnets can exceed this limit many times over.

This was the beginning of a new era never examined before. A plethora of applications of superconductive metals and alloys has been envisioned and is currently under research and development. These are in the area of electric power generation and transmission, in high energy particle accelerators, in magneto-hydrodynamics (the science dealing with interaction of a magnetic field and a conducting fluid such as ionized gas, liquid metal, etc.), in microwaves, in nuclear fusion systems, in vehicular transportation (such as magnetic levitation train or magneplane, etc.) and others.

"Science is an imaginative adventure of the mind seeking truth in a world of mystery."

—Sir Cyril Hinshelwood (1897-1967)
English Chemist, Nobel Prize Winner 1956

Celestial Bodies

"For most of human history we have searched for our place in the cosmos. Who are we? We find that we live on an insignificant planet of a humdrum star lost in a galaxy tucked away in some forgotten corner of a universe in which there are far more galaxies than people."

— Dr. Carl Edward Sagan (1934-1996)
American Astronomer, Searching for Intelligent Life in the Cosmos

Celestial bodies (also called Heavenly bodies) are the huge aggregates and groupings of the matter particles, defined as a collective unit of mass in the heavens or the firmaments. Celestial bodies include things such as moons, planets, stars, suns, asteroids, comets, nebulae, etc. One such aggregate as a whole is capable of exerting large forces on other bodies and reversely be influenced by comparable or larger bodies.

STAR SYSTEMS are **Very Large Macroscopic Matter** where the planets circle the stars in a relatively circular or oblong fashion to keep the created space un-collapsed and thus, stretched out. There exist large interplanetary forces (between heavenly bodies), which is the summation of many small forces within each heavenly body.

Therefore, a body's motion in a star system (such as earth in the solar system), is governed primarily by all the forces exerted by the heavenly bodies (near and far). For example, earth is pinned (or held) in place in a relatively fixed orbit by gigantic forces coming from the sun and other large planets in the solar system. This system appears to be in a balanced and unchanging state of affairs on a large scale. However, at each moment of time, there are changes, even if very small, that exist in many levels but are difficult to observe. This is because one is interior to the solar system. Thus, any observation that is done at this moment is purely secondary and is performed through astronomical instruments such as powerful optical devices and radio telescopes.

Galaxies are **Ultra-Large Macroscopic Matter** where the groupings of aggregates of masses of stars, planets, and asteroids form a series of spirally shaped (or elliptically shaped, etc.) energy and matter, spread almost on a plane with a hub and containing thousands of stars each surrounded by planets, asteroids, nebulae, and celestial gases. One of these spirally shaped masses is called a galaxy. The position of each galaxy changes relative to other galaxies. This is due to the existence of huge and unfathomable forces that galaxies mutually exert on each other.

CLUSTERS OF GALAXIES are **Super-Large Macroscopic Matter** where the galaxies aggregate into huge masses called "clusters of galaxies."

"Shoot for the moon. Even if you miss it, you will land among the stars."

— Les Brown (1912-2000)
American Songwriter

Matter and Energy

"The energy of the mind is the essence of life."

—Aristotle (384 - 322 BC)
Greek Philosopher and Scientist

Before 1905, mass and energy were considered as separate entities and the principle of "*conservation of energy*" was often supplemented by the principle of "*conservation of mass*," however, with the advent of the theory of relativity proposed by Albert Einstein, ***mass and energy*** were discovered to be *interchangeable* and *equivalent* to each other.

In other words, the origin of matter is energy. Energy could be converted into mass by reducing the volume of the energy particles (such as turning steam into water or ice by cooling it down and shrinking the space between the water particles), and vice versa, mass could be converted into energy by nuclear fission or fusion, which decompresses or decomposes matter particles (atoms) into the constituent energy particles (electrons, protons and neutrons).

Therefore, the principle of conservation of mass became a special case (or a subset) of the principle of conservation of energy, and a more general principle was born: *Mass-energy may not be created or destroyed, but may be converted into each other.*

"All the breaks you need in life wait within your imagination. Imagination is the workshop of your mind, capable of turning mind energy into accomplishment and wealth."

—Napoleon Hill (1883-1970)
American Author

Relative Size of Particles

Decomposing an atom into its constituent components leads to several important particles. These particles have been described in depth in this section, chief amongst them are the following with their relative sizes approximated to the nearest powers of ten:

Atom~ 10^{-8} cm
Nucleus~ 10^{-12} cm
Proton/Neutron~ 10^{-13} cm
Electron< 10^{-16} cm
Quark<10^{-16} cm

"The scientist only imposes two things, namely truth and sincerity, imposes them upon himself and upon other scientists."

—Erwin Schrodinger (1887-1961)
Austrian Quantum Physicist

These particles with their relative sizes are shown below.

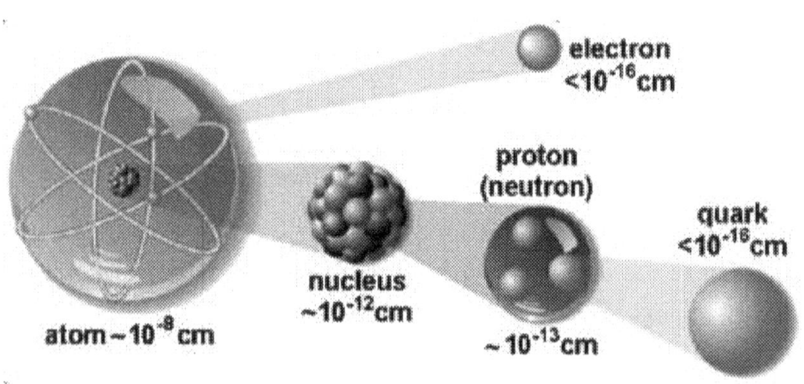

The Relative Size Of Matter Particles.

PART VIII

The Phenomena of Time

Time

"Don't wait. The time will never be just right."

—Napoleon Hill (1883-1970)
American Author

Time is a consideration, which is a result of matter and energy in action and actually proceeds from their interactions. We know that space precedes the existence of matter and energy; however, time proceeds from them.

The constant and continuous interaction of matter and energy with each other brings about the concept of time.

"All that really belongs to us is time; even he who has nothing else has that."

—Baltasar Gracian (1601-1658)
Spanish Philosopher and Writer

Objective Time

It is essential to note that by time most people mean **objective time** (also called mechanical time), which is linear. It is a linear pseudo-commodity which cannot be exchanged like the other three commodities of created space, created energy, and created matter.

By the term "linear" we mean something that has the same amount of change for each exact interval. In other words, the term "linear" means "uniform rate of change." For example, if the mass of a wall increases 20 lbs per foot throughout its entire length (i.e., it has a uniform density of 20 lbs/ft), then that wall is said to be linear. On the other hand, if part of a wall has a uniform density of 20 lbs per foot while another part of it has a density of 50 lbs per foot, then that wall is said to be nonlinear.

The basis of mechanical time originates in altering the position of a particle in a region of space relative to another particle. Thus, in order to define time, the following requirements need to be met:

a) First, a region of linear space needs to be created, and
b) Second, matter particles (or objects) are to be placed in this region of space,
c) Third, these matter particles must have sufficient energy to cause their motion relative to one another.

With this preamble, "time" comes into its own, and is defined as "*A measurable extent of change—change of position of a particle relative to a starting point in a created space.*" It is a primary postulate used in the creation of the physical universe (on a macroscopic and/or microscopic level) and its value at any given location orders the sequence of events at that point in space.

One way of thinking about time is to imagine a world that is standing still, which would be timeless. Now if we allow some type of irreversible change, then that timeless world would be different "now" than what it was "before." The period between the present moment (now) and the past moment (then) is called time, which can be measured and recorded as an index of change.

Thus, we can see that the concept of time, as a very general concept, actually embraces the concepts of "change" and "measurement of such a change." Time proceeds from the interaction of matter and energy located in a certain space under consideration, and its measured value is purely used as an ordering index of sequential events, thus, by itself it is a great organizing tool. It is used to keep track of the change of a moving particle's location, very useful for bookkeeping purposes.

"There was no "before" the beginning of our universe, because once upon a time there was no time."

—John D. Barrow (1952-)
Professor of Physics, Cambridge University

The fundamental unit of time measurement is supplied by the earth's rotation on its axis while orbiting around the sun. Earth's two prime motions are a) rotation on its axis, and b) orbiting the sun as shown below. Since these two rotation motions are both at fixed speeds, therefore, they create a linear scale on which we can measure change. Therefore, mechanical time is linear as a result of linearity of earth's motions.

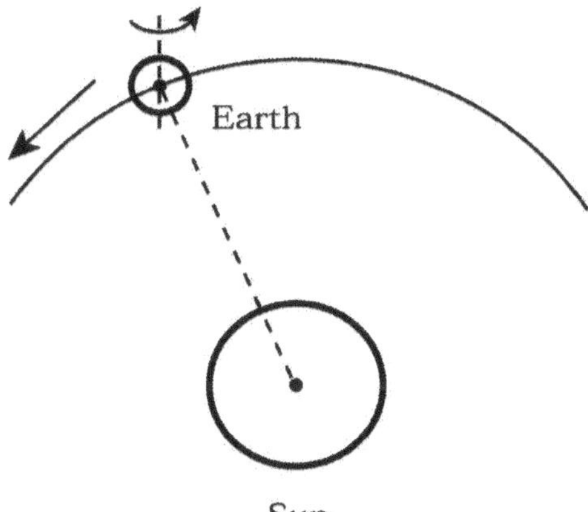

Earth's Two Prime Motions Supplying Fundamental Units Of Time.

So time is the action of energy in space and its keynote concept is, **“Change in location relative to.”** Thus, to have time, the particle’s location or its relationship in space should change relative to its starting point, ending point, and to other particles or objects.

"Nothing is as far away as one minute ago."

—Jim Bishop (1907-1987)
American Writer

As stated earlier, time measurement is basically a method of bookkeeping with regard to change of a particle relative to its starting point. For example constant change of location of earth around itself as well as relative to the sun and other planets takes place at a certain and fixed rate and gives us mechanical time which can be measured electronically on a digital clock or mechanically by a watch or an hourglass, etc. This revolving motion, which runs on an automatic basis, has been well analyzed and dissected, and has been arbitrarily adopted to provide us the “measuring scale” against which all other motions that we observe in the physical universe are measured.

The selection of earth’s motion for the purpose of defining the fundamental units of time (whether in seconds, minutes, hours, days, etc.) is totally an arbitrary factor introduced primarily for convenience. Other linear motions could have very well been used, and therefore, another series of agreements regarding the time base (such as the motion of the moon or Mars or Jupiter around the sun) could have been made. Furthermore, it could have been any other totally arbitrary measurement scale such as 1 time-unit=1.654 seconds, etc. Here is the first clue about the dependency of “time” on, “The considerations of the viewpoint,” and how the viewpoint affects its units, its measurement methodology and its final interpretation.

"The distinction between past, present, and future is only a stubbornly persistent illusion."

—Albert Einstein (1879-1955)
German Born American Physicist, Nobel Prize in 1921

Subjective Time

There is also the consideration of time in one's mind, which could be called the **subjective time**, and is a nonlinear or linear quantity depending on one's viewpoint.

So, the concept of time is a double-edged sword. At first glance, time appears to be a simple concept of looking at our watch and telling what its reading says. But this is only a read-out of something that is occurring on a "continual change" basis.

The measurement and recording of this thing which is occurring on a "continual change" basis is done by two methods:

a) The first method is through the use of mechanical measurement and recording techniques such as clocks, time registering machines, videotapes, etc. This we call, "Objective time or mechanical time," which is linear, and
b) The second method of measurement and recordation is done on a mental level and is registered as an experience, with a "Subjective time" index along side of it. Subjective time measurements or recordings can become linear or nonlinear depending on the viewpoint's considerations.

The two concepts of objective time and subjective time are shown on the next page, where both are subject to viewpoint's considerations.

"Time does not always pass at the same speed. We are the ones who determine that speed."

—Paulo Coelho (1947-)
Brazil's Mystical Author

An example may elucidate this point further. Let us consider a viewpoint experiencing unforeseen conditions, which has caused tremendous mental stress such as an accident. For this viewpoint, the subjective time is measured and recorded highly nonlinear, even though the objective time recording remains very linear!

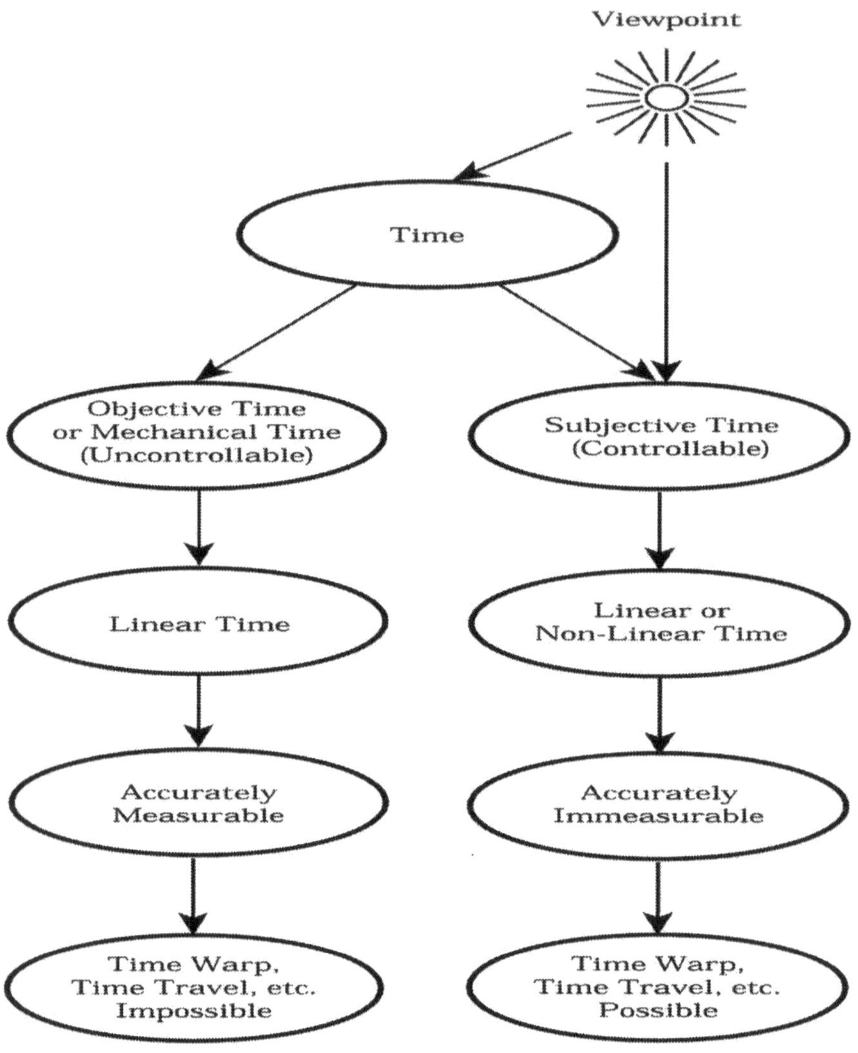

The Concepts Of Objective Time And Subjective Time.

This is only one example; there are many other instances that one can think where subjective time becomes very nonlinear (see the Figure on the next page). From this figure, we can see that such things as, "problems, accidents, etc.," speed up the subjective time, whereas events such as, "Waiting, etc.," slow time down. However, the mechanical time remains completely linear during these nonlinearities in the subjective time domain.

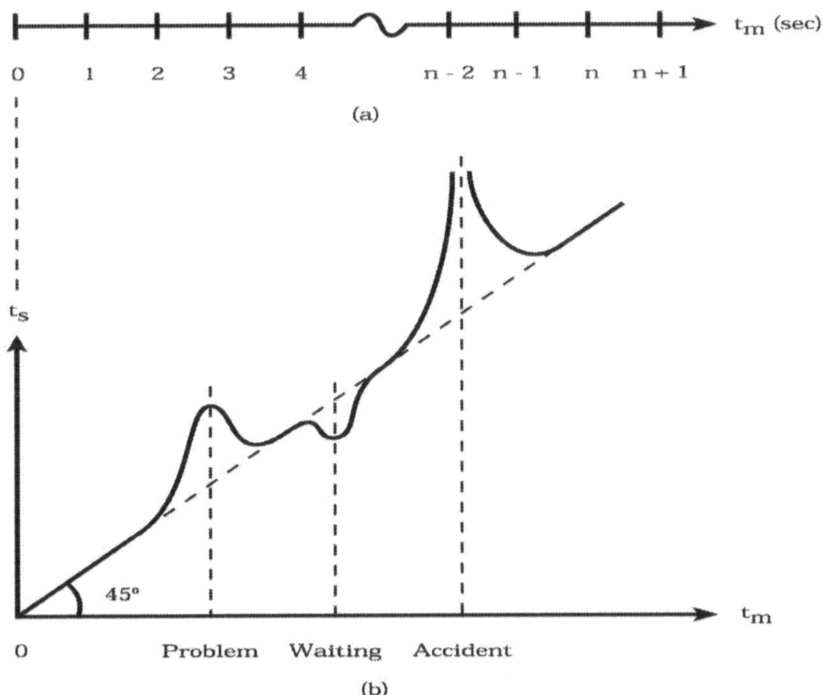

a) A Linear Time Axis, And b) Subjective Time (t_s) *Vs. Mechanical Time* (t_m).

Therefore time, even though measurable on an objective basis, has a subjective side to it, which is an abstract concept representing "change".

"Time is too slow for those who wait, too swift for those who fear, too long for those who grieve, too short for those who rejoice, but for those who love, time is eternity."

—Henry Van Dyke (1852-1933)
American short-story Writer, Poet and Essayist

Measurement of Time

Time represents change and any change that repeats itself stands out from other irregular events. For example, the rising and setting of the sun are examples of *natural repeating change*, which were originally used by the first people to keep track of non-repeating events such as, a simple action of meeting a new person, or going on a trip to see a new location for the first time, a natural catastrophe, etc. Later, Man imitated the regularity of natural events by inventing different types of clocks.

Nowadays, with the help of very sophisticated instruments, we can measure time very accurately and regularly, on a short or long term basis.

With this preamble about time, we can see that the measurement scale required to measure objective time is provided by objects or bodies that move at a fixed and known rate, such as the earth orbiting the sun or, better yet, the nine planets as a collective mass in the solar system orbiting the sun at a relatively constant speed *(natural repeating changes)*. Existence of time can be subdivided into four strata:

a) Time at first was a postulate, "*Let particles of energy and matter change their location.*"

b) Then it manifested itself throughout the physical universe as change on a microscopic and macroscopic scale.

c) Next, we have a measurement system (such as clocks) set up at each location to measure this change.

d) Finally, we have a value or time-stamp (such as 2:45 pm in Los Angeles) generated by the measuring system, which we now call, "time." So, the concept of time has trickled down through several strata of interpretation and substitutions to arrive where it is now—a numerical digit on our watch dial!

The figure below summarizes and shows the relationship between these four strata of time.

Objective time can be measured in the physical universe very accurately within a fraction of a second. Each segment of it can have a beginning and an exact ending point. On the other hand, subjective time is one continuous stream of change with no specific beginning or end and only gets a time stamp from the objective time side of existence.

"There is never enough time, unless you're serving it."

—Malcolm S. Forbes (1919-1990)
American Publisher and Editor

Actuality of Time	Postulate of Time	Stratum 1
Manifestation of Time Postulate	Change in the Physical Universe	Stratum 2
Use of Stars as a Reference for Time Measurement	Time Measurement System	Stratum 3
We Refer to the Read-Out Values as Time	Clock Read-Out	Stratum 4

The Subdivision Of Time Into Four Strata.

Time as a Fourth Dimension

"I like the dreams of the future better than the history of the past."

—Thomas Jefferson (1762-1825)
American 3rd US President (1801-09),
Author of the Declaration of Independence

There is a considerable discussion on time as a fourth dimension amongst physicists. They see the change taking place in a three dimensional created space of the physical universe and they want to plot this change as a fourth dimension of the physical universe.

There are several technical problems associated with this action. First of all, time and space are not cut from the same cloth, so to speak. Space is senior to energy and matter. Since time is a result of their interaction—a derived and secondary byproduct of this interaction, therefore it is junior in importance relative to space. Thus, space and time are not of comparable magnitudes!

Secondly, Time in the strictest sense of the word, is a postulate about change. However, actual change in the physical universe as represented as, "*time in physics texts*," is the second stratum, which is a cut-down from this postulate. One should recall, before there was change there was the postulate of time; therefore, the two strata are not the same thing.

Also, as a counter argument, we can see that the concept of fourth dimension is dependent on the change of a particle's position, or the interaction of matter particles at each point in space. However, if there are no matter and energy to interact at that point in space, then there is no change, thus, the fourth dimension instantly disappears!

In other words, space can exist without any matter or energy in it, thus, it would have no time associated with it, and it would be timeless. Therefore, time could possibly be considered a fourth pseudo-dimension, at best.

As a method of bookkeeping and keeping track of "change" of position of matter and energy particles at a locale, it would be alright to plot the change at that location (i.e. time) alongside the three dimensions of space as shown on the next page.

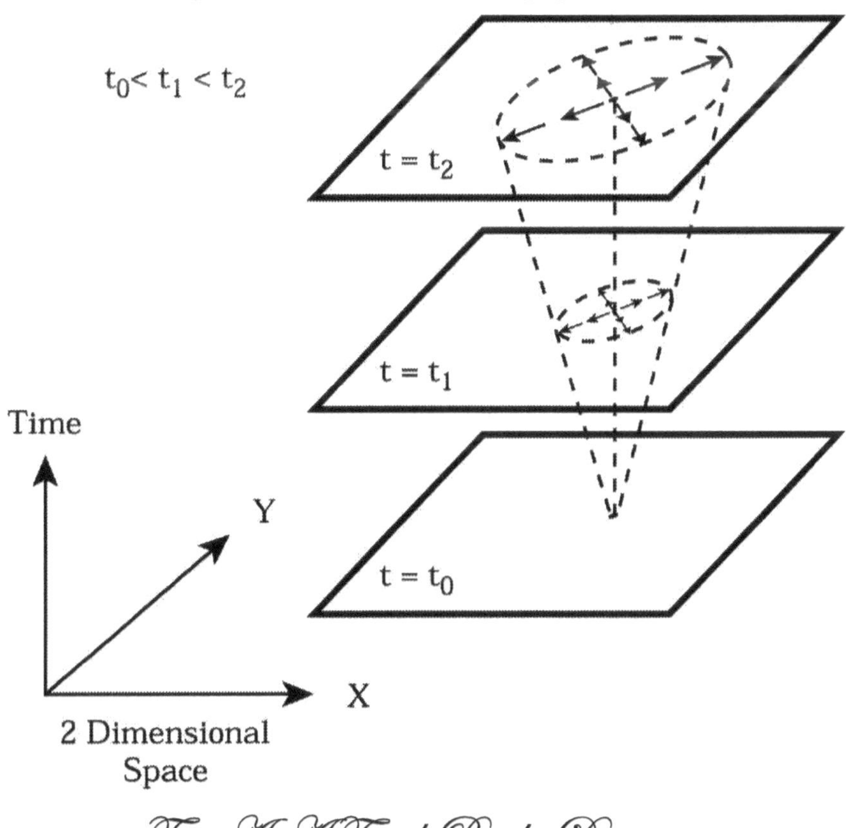

Time As A Fourth Pseudo-Dimension.

"He who spends time regretting the past loses the present and risks the future."

Quevedo (1580-1645)
Spanish Master Satirist and Poet of Spain's Golden Age

However, this plot of change, visually elegant could not warrant the addition of a new dimension, because it is not on the same order of magnitude with space. It is simply an apparency and not an actuality. In other words, it appears to be a fourth dimension and in actuality it is not!

Mathematically, the three variables of space are of the same order of importance as that of the fourth variable, even though in actuality it is not a correct assumption.

Such a concept, as delineated above, is referred to as "space-time" by physicists. Their main assumption is that all space is always occupied by matter or energy. Such an assumption, even though most usually encountered in practice, is fundamentally incorrect. This is because once space is created, the resultant creation (called created space or simply space) can exist independent of mass and energy, and as long as it is kept as a created space, it will behave completely linear in a timeless fashion.

"An unhurried sense of time is in itself a form of wealth."

—Bonnie Friedman (circa 1965-)
American Writer

Time Warps and Time Travel

"Your future takes precedence over your past. Focus on your future, rather than on the past."

—Gary Ryan Blair (circa 1965-)
American Motivational Speaker and Author

Similar to space warps, there are time warps and time travel, which is when one travels to the past or visits the future and then comes back to present time. These time discontinuities or warps can be elegantly explained by pure mathematics and even simulated electronically, graphed and displayed by advanced software. However, that is where the time warp ends.

In actual practice, the mechanical time created by the motion of stars, planets and galaxies cannot be reversed. Even if it could be reversed, it would be a new time track and would not be equivalent to delving into the past. This is so because the particles, objects and bodies that formed the past moment are no longer at that exact location and even if all planets could be gathered at that exact same configuration, it would still form a whole new time track and would not make, for instance, a tea cup falling down into many pieces, jump back up in reverse order and become whole again.

Moreover, in order to have a time warp, we need to create a time discontinuity. However, since existence of time is totally dependent upon the existence of space, energy, and matter; therefore we need to have first and foremost a discontinuity in space, which is the most senior element. Then on a secondary level, energy has to be discontinuous, and then finally, matter has to be discontinuous. The discontinuity of all of these three would eventually lead to the existence of a time warp.

The illusion of reversing time is created by playing backwards an incident, which has been recorded on a storing device such as a videotape. This is the world of movies, illusions, fiction, and fantasy where we have magic, space warp, time travel, and supernatural events. Time discontinuities, time warp, and time travel are not part of the reality of the situation or the actuality of the physical universe.

The make-up of the physical universe does not allow mechanical time travel into the past or future. It always *exists and operates at the present time*; even if one views a recording of a past event, he still views it in the present time.

However, if we consider the world of mathematics, we can see that time warps and time travel are quite possible. They become workable concepts in mathematics, which is a highly abstract and idealized subject. The reality of our physical universe is quite something else.

On the other hand, reliving a past experience mentally or imagining a whole new time stream for some future event would be equivalent to time travel in the "subjective time" stream, which is a reality and quite possible. Subjective time being highly nonlinear and at the command and under the consideration of the viewpoint is highly susceptible to time warps and time travel and other time oddities. However, that information is quite subjective and not visible to the naked eye and thus, becomes impossible to demonstrate in the physical universe.

In the final analysis, we could consider two types of time a) Objective or mechanical time, which is linear and easily measurable by a mechanical gadget such as a clock, and b) Subjective time, which deals with the experience and recordation of "change" or "motion." However, since the concept of "change" or "motion" depends on how the viewpoint sets up the space to view the change or motion; therefore, time in the strictest sense of the word is a "consideration about change and/or motion." In other words, time can be a linear or nonlinear pseudo-commodity, dependent on the viewpoint's consideration!

"Time does not change us. It just unfolds us."

—Max Frisch (1911-1991)
Swiss Novelist and Writer

The Universal Time Base

The physical universe on a rigorous level does not allow a "mechanical time shift or time discontinuity" of any kind. However, a time shift can occur in the physical universe purely by an agreement between all parties involved. For example, twice a year in the United States, there is a one hour time shift in most states. This time shift occurs every fall and every spring season, and the local time is shifted by a prior agreement, which was made sometimes in the past on a state level.

Therefore, the notion of the actual time, which is the local time is set aside due to some consideration concerning daylight hours. This example clearly shows that time is basically a consideration at its root, even though it manifests itself as an imposing number on our watch dial.

Another example is the different time values at different locations on earth such as 4:00 o'clock in New York corresponding to 1:00 o'clock in Los Angeles. This at first may not seem to pose any problem, however, traveling between different locales with different time zones will bring about time shifts, with a resulting time mix-up in one's mind.

"Anger blows out the lamp of the mind."

—Robert G. Ingersoll (1833-1899)
American Statesman and Orator

A possible solution to this time-shift problem and the time disarray that it creates for the earth's inhabitants, is using a "new time base" for time measurement. The current time base is tied to the earth's spin around itself and its rotation around the sun. However, the new time base must move away from the "earth's spin and rotation around the sun" considerations and establish a Universal Time Base (UTB), which has only *one value all over the world.*

To implement the Universal Time Base, a system could be set up based on a decimal number system, such as ranging from zero to one hundred, to measure the seconds, minutes and hours in a day separated by a comma or a slash (e.g., 100,100,100,…), so on and so forth. There should be conversion formulas and conversion tables so that one can convert from the current Earth Time Base (ETB) system of measurement to UTB system. In this manner, all of the current aspects of time measurement in terms of seconds, minutes, hours, days, months, years, centuries, and millennia (such as the Roman calendar, etc.) could be accurately accounted for in the UTB system.

The zero point of time is an arbitrary factor and could be taken as any mutually agreed upon point in time, by all scientists. It is actually immaterial for accurate time measurement, but is a needed factor for a practical and useful system.

One of the biggest advantages of the UTB measurement system, aside from decreasing the time confusion on earth, is preparing Man for space travel.

At the current rate of technological development in vehicular motion and new sources of energy for outer space travel, Man is embarking on a whole new track of existence since his appearance not more than 100,000 years ago. Interplanetary and intergalactic travel is not in too distant a future, where the concept of "earth around the sun" and ETB become meaningless and obsolete and thus, a more advanced time base is needed.

ETB generates time outputs, which appear to be "circular" such as the 2:00 o'clock hours in consecutive days, Mondays in consecutive weeks, July months in consecutive years, etc. Thus, the UTB system may also be used to prevent the "circular" notion of time brought about by the motion of earth around itself as well as the sun and the ETB system of time measurement.

"Time, whose tooth gnaws away at everything else, is powerless against truth."

—Thomas H. Huxley (1825-1895)
English Biologist

A Time Oddity

There is a time oddity, which should be discussed in more depth since its existence has introduced a falsehood in the subject of time. This oddity can be illustrated by an example. If one travels from New York to Los Angeles, one gains three hours. Furthermore, if he continued his trip from Los Angeles to Tokyo and through London back to New York, it appears that he is "gaining time" and as a result is "getting younger," or is "traveling into the past," which is an apparency.

Reversing the trip in this example (i.e., traveling from Los Angeles to New York, to London, to Tokyo, and back to Los Angeles), one appears to be "losing time," which may be interpreted as "getting older" or "traveling into the future." This is also an apparency!

Both of these apparencies are falsehoods, which are brought about by the utilization of the various local time zones and the earth based time (ETB) measurement system, giving the erroneous concept that one is gaining or losing time, which may be construed as, "Time traveling into the past or future in the physical universe."

The actuality is that the physical universe is vibrating at a certain rate, which automatically creates mechanical time. When one moves in it in any direction regardless of the number of time zones one crosses, one has caused mechanical time to elapse.

"Losers live in the past. Winners learn from the past and enjoy working in the present toward the future."

—Denis Waitley (1933-)
American Motivational Speaker and Author of Self-Help Books

Moreover, since any and all part of the physical universe *always operates and exists at present time*, thus, arriving at any point on earth corresponds to no actual time gain or loss but only an apparent and previously-agreed-upon time shift to the local time zone.

This time oddity, as innocuous as it may seem at the moment, has caused and will continue to create time confusions particularly for space travelers, and thus, may necessitate the introduction or even hasten the rapid implementation of a time measurement system based upon the Universal Time Base (UTB), which is independent of where one is on earth (or any other planet for that matter). UTB is neither susceptible to the introduction of time oddities and falsehoods into the time measurement system nor is it prone to erroneous concepts of time travel (going into the past or the future), which is impossible in the physical universe.

In conclusion, the UTB system of time measurement avoids all of the time oddities of the ETB system (such as local time shifts, time circularity, etc.) and provides a highly accurate indication of mechanical time, which is a monotonically increasing and yet linear index of change of a particle's position in space.

"Nothing is a waste of time if you use the experience wisely."

—Auguste Rodin (1840-1917)
French Sculptor

Traveling at the Speed of Light

As velocity (v) of an object approaches the velocity of light (c), its mass approaches infinity. In other words, apparent mass of an object is a *variable quantity* depending on its speed, and since an infinite mass is an absolute and thus an impossibility in the physical universe, therefore the speed of light becomes the limiting speed of all masses.

The increase of mass of a moving object is not due to a magical increase in the number of the atoms comprising the object, but is an apparency caused by the increased momentum. It is like a bullet, which has more mass when moving than when standing at rest.

"***There are no speed limits on the road to excellence.***"

—Unknown

Furthermore, as the speed of a particle approaches the speed of light, the time associated with that particle (t) starts to slow down toward zero. This phenomenon was first discovered mathematically by Lorentz and Fitzgerald and then later interpreted by Einstein for proper application to the physical universe, as shown in the table on the next page.

Using this table and considering the current speed of the fastest aircraft to be Mach 3 (triple the speed of sound, 3x344 m/s equal to 3x1129 ft/s or 3x0.213 miles/s) or approximately 2,300 miles per hour or a mere 0.64 miles per second (cf. 186,206 miles/sec, the speed of light), we shockingly realize that when we speak of time shift we are dealing with a totally abstract concept, a mathematical concept existing in the theoretical world, which is far beyond the realm of reality and is impossible to prove or disprove at this stage of man's technological development.

v (miles/sec)	v/c	m/m_0	t/t_0
0	0	1	1
0.213 (sound)	1.145×10^{-6}	1	1
46,552	0.25	1.032	0.969
93,103	0.50	1.155	0.866
139,655	0.75	1.512	0.661
167,586	0.90	2.294	0.436
186,187	0.9999	70.712	0.014
186,206	1	∞	0

A Tabulation Of Mass And Time Versus Speed

(m_0 and t_0 are the mass and the time measured at rest)

It is also understood that to achieve such a high speed (as the speed of light), the observer can not be in a human body form, but only can exist as a body-less entity (or a viewpoint) capable of pure observation. Such a body-less observing viewpoint exists only on a theoretical plane at this moment in the development of sciences.

"Pain reaches the heart with electrical speed, but truth moves to the heart as slowly as a glacier."

—Barbara Kingsolver (1955-)
American Writer

PART IX

Dichotomy and Duality in the Universe

The Dichotomy Principle

English dictionary defines "***Dichotomy***" as *"two things or concepts that are sharply or distinguishably opposite to each other."* On a broader look, we can see that the concept of dichotomy is an old concept in philosophy but sciences have given a new meaning and definition to it and have used it very effectively to facilitate the understanding and organization of their body of knowledge.

In actuality we can observe that the concept of dichotomy is based upon a much more fundamental truth, and that is the "***Principle of Relativity of Knowledge,"*** which states that *all derived knowledge is relatively true*.

From this principle we can see that a datum cannot exist all by itself (i.e., an absolute) and when plotted on a linear scale from 0 to +∞ is surrounded perforce by many comparable-sized data. On the opposite side, we have a scale extending from -∞ to 0 (matching up with the 0 to +∞ scale) wherein lies the equal and contrary datum. Examples of dichotomies abound in nature and include such things as "***Kinetic-Static,***" "***Cause-Effect,***" "***Create-Destroy,***" "***Something-Nothing***," etc.

By observation of the materials that the nature, or in general the physical universe have presented to us, we can define "***The Dichotomy Principle"*** as "*Materials and concepts connected with the physical universe, such as laws, principles, quantities, etc., coexist in opposite pairs.* "

"Youth has no age."

—Pablo Picasso (1881-1973)
Spanish Artist and Painter

⊶ ꕥ ✶ ❁ ✶ ꕥ ⊶

Static vs. Kinetic

Through a close observation of the physical universe we can see that every physical object or entity is composed of two parts:

a) *The consideration component, which is a "static" and unchanging, and*
b) *The actual physical component, which is a "kinetic," meaning that it is moving constantly microscopically (on an atomic level), even if not macroscopically (on a large scale).*

Thus we see that "***Static and Kinetic***" form a neat pair in dichotomy.

By static we mean something which is truly without motion or change, such as truth or a postulate. In physics, one may consider a very distant star (a physical universe object) a static on a short term basis; however, this is not totally correct because the distant star moves over a long period of time, thus is not truly a static in the strictest sense of the word, but only an approximation or a physical analogue of a true static in the thought universe.

Needless to say that anything motionless on earth (such as a building, a bridge, etc.) is not a true static but actually a kinetic, due to earth's constant motion in several directions.

"In the end, we will remember not the words of our enemies, but the silence of our friends."

—Martin Luther King, Jr. (1929-1968)
American Baptist Minister and Civil-Rights Leader

The Duality Principle

"Heart knows neither duality nor the limitations of space and time."

—Sri Sathya Sai Baba (1929-)
Indian Spiritual leader

An important *subclass of dichotomies* is the concept of "***duality,***" which has special significance in sciences and thus needs to be addressed in depth.

The word ***"Dual"*** comes from the Latin word ***"Duo,"*** which means "two." In scientific jargon, the word "dual" is used to mean two concepts or things that are comparable to each other but are opposite in nature and thus act as counterparts to each other.

The items in a pair of dual concepts complement each other and therefore form a complete set. For example, concepts of (positive & negative), (electron & hole), ($+\infty$ & $-\infty$), are pairs of dual concepts. Furthermore, combining the concept of duality with the principle of relativity of knowledge gives us the whole panoramic view of our physical universe in its entirety and covers all bases as shown on the next page.

On a more general level, we can see that every material item in the physical universe is matched up with its dual, and together they create a locked up standing wave, a sort of equilibrium of forces which has a location but does not allow a real indefinite flow to exist and thus causes the physical universe, as a whole, to float in time. The existence of life force may cause an imbalance of this equilibrium of forces and thus create a momentary flow of energy, but it is against a heavy backdrop of this major standing wave, which exists mechanically and heavily influences our daily existence.

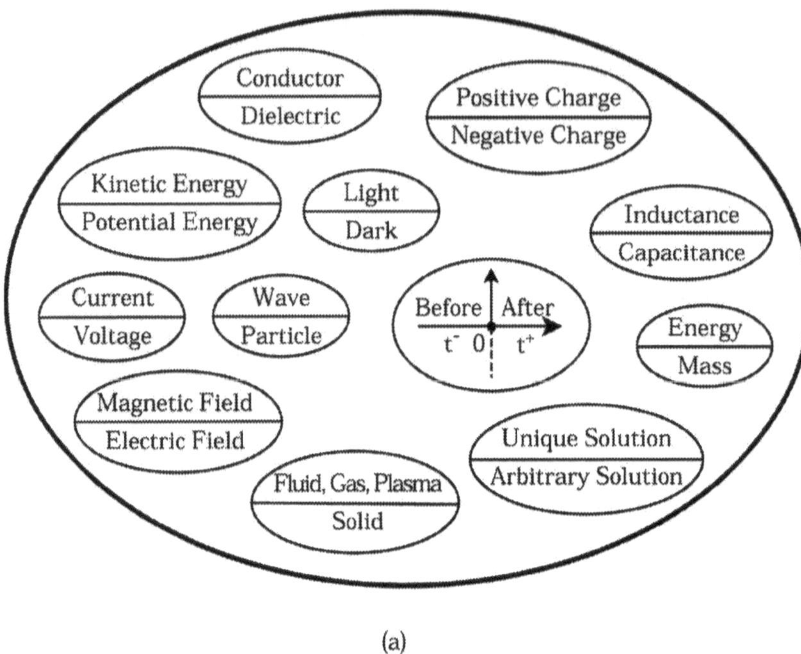

(a)

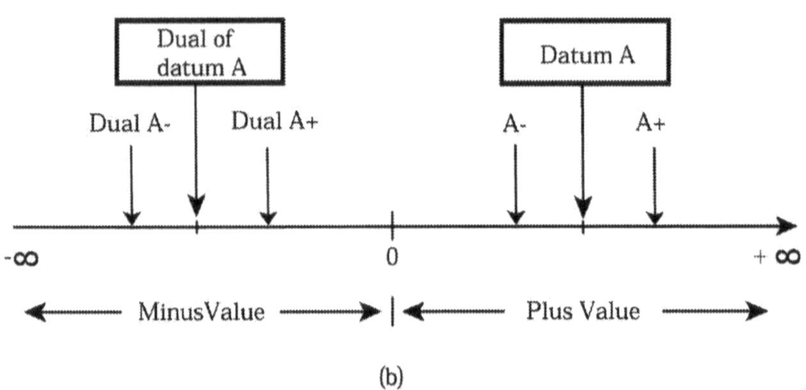

(b)

a) Examples Of Dual Concepts, and b) Duality Combined With The Principle Of Relativity Of Knowledge.

For example, considering the field of electricity, we can see that the only form of electricity, prior to Volta (1800 AD), was electrostatic charge, which is a standing wave. It took a considerable amount of work on the part of the life force of Man as a species, to discover and produce a real flow of electricity (as in a closed circuit energized by a battery) and thus create a minor imbalance in this standing wave type universe.

"Act with kindness, but do not expect gratitude"

—Confucius (551-479 BC)
China's Most Famous Teacher, Philosopher, and Political Theorist

The concept of **"Dual"** is used extensively in engineering and when used in this book we mean specifically, *two concepts, energy forms or physical things that are of comparable magnitudes but of opposite nature, thus becoming counterpart of each other.*

The concept of "dual" leads to the ***"Duality Principle,"* (*or The Principle of Duality*)** which is a very important concept in sciences and engineering and is defined as "*The principle that for any theorem there is a dual theorem, in which one replaces quantities and operations with their duals. In other words, if a theorem is true, it remains true if each quantity and operation is replaced by its dual quantity and operation."*

By observing and understanding one of the items of a pair, the properties of the other item in the pair can be predicted and grasped. Thus one can gain a deeper understanding about each item or the whole pair by this process. .

In the previous section we discussed the dichotomy of "*Static and Kinetic*." Care must be taken here at this point because these two concepts (i.e., static and kinetic), even though a pair in dichotomy, are not of comparable magnitude or nature, thus are not dual of each other.

"Every explicit duality is an implicit unity."

—Alan Watts (1915-1973)
American Writer

Discovery of Unknowns

Dealing with new territories or unknown fields of study, it is usually possible for scientists or researchers to apply the principle of duality to "unknown fields" with great success. They can gain much understanding by obtaining solutions for the "unknown field" by using the results from the "known field" through replacing the dual quantities and operations in the "known solutions" and thus know about an unknown field far in advance of its actual discovery.

Utilizing this powerful principle, one can gain a deeper understanding about an "unknown," whether a datum or a subject than usually is possible, provided one knows the known dual.

"The difference between what the most and the least learned people know is inexpressibly trivial in relation to that which is unknown."

—Albert Einstein (1879-1955)
German Born American Physicist, Nobel Prize in 1921

As an example, James Clerk Maxwell used the duality principle in predicting the relation of "current flow and magnetism" in the field of electricity, which was an unknown at the time. He knew "electricity could create magnetism," so he proposed its dual, which is "magnetism should create electricity." This prediction was well in advance of its actual discovery, but he discovered the unknown by using the duality principle.

Furthermore, by using this principle in his mathematical work, Maxwell predicted the existence of waves over twenty years in advance of their actual proof through experimental means, by a German scientist named Heinrich Hertz.

"There are things known and there are things unknown, and in between are the doors of perception."

—Aldous Huxley (1894-1963)
English Novelist and Critic

Analysis and Design

"If you don't design your own life plan, chances are you'll fall into someone else's plan. And guess what they have planned for you? Not much."

—Jim Rohn (1930-)
American Motivational Speaker, Philosopher and Author

We are living in a technical world dictated by the principles of physics and engineering, so let us define "**Engineering**" as, "*The science concerned with putting scientific knowledge to practical uses, which includes the planning, designing, construction of machinery, equipment, etc. and their proper placement in the physical universe.*"

In other words, Engineering is the science by which properties of matter and the sources of power in nature are explored and made useful to life forms in terms of structures, machines, and products.

Using this discussion and on a broader scale of observation we can see that engineering sciences are primarily concerned with either "analysis" or "design," which we need to define fully at this point:

By "**Analysis,**" we mean breaking up of any complex problem, machinery, network, and so on, into its simple elements or its constituent components, wherein through an examination of these parts, their nature, function, interrelations, frailties, etc., can be discovered toward an end result of understanding the mechanics of the operation as a whole.

By **"Design,"** we mean the act of conceiving a pattern and planning a structural layout as well as determining specific parameter values (such as voltage, current, power consumption, etc.) for the construction of any desired system or process, which can be visualized. The end result of the design process is the creation of application mass such as a device, a circuit, a system, an experimental setup, a sequence of related actions, etc.

Upon placing these two distinct areas in a proper perspective, it becomes evident that analysis or design in engineering can be summed up as:

"**Engineering Analysis**" of a desired object is dealing with and taking apart its created space, created energy, created mass, and created time stream. On the other hand, "**Engineering Design**" brings the dynamic factors of "creation of space, energy, mass, and time" into focus. **Therefore, we can observe that from the point of view of engineering, "Analysis and Design" are concepts in dichotomy**.

"Happiness is not something you postpone for the future; it is something you design for the present."

—Jim Rohn (1930-)
American Motivational Speaker, Philosopher and Author

When one is troubleshooting a device such as a TV, one is dealing with the created space (actual dimensions, circuit layout, etc.), the created energy (specific voltage and current values), the created mass (actual bulk and sheer mass of the many parts), and created time stream (manufacture date) of the TV. This is what we usually refer to as analysis.

On the other hand, from the first moment of designing a TV, one is embarking upon the direction of creating space, energy, mass and time for an intended object. This actually means that the paper design, the simulation, and the component layout stages are the "create space" stage, whereas the decision about the voltage, current, electric, and magnetic fields and its final implementation falls in the category of creating energy. The actual assembly of all of the components and parts, which would make a TV functional, would fall in the category of "create the mass" stage of the TV. Finally, exactly at the inception point when the TV is rolling off the assembly line, a time stream is created for that particular TV. This time stream is totally new and has never existed before.

On a simpler level, we can see that engineering analysis is firmly founded on the principles of the physical universe and takes its cue from the science of physics but goes beyond it in the direction of a whole new field of creation, which is dealt with in the field of engineering design.

Therefore, we can observe that **Physics** only explores "***The created space, energy, mass, and time***" side of existence, which is really just one side of the "coin," so to speak. The opposite side of the coin, **Engineering**, would deal with "***Creating space, energy, mass and time,***" which is highly viewpoint dependent but is a much more dynamic subject, as shown below.

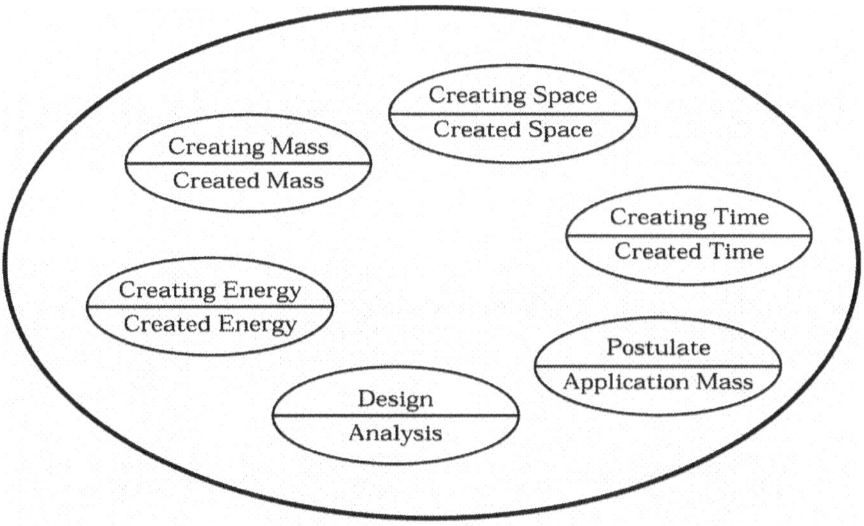

The Dichotomy Between Creating and Created, or Physics and Engineering.

It is interesting to note that there are other subjects other than engineering that deal with creation of space, energy, mass and time. These subjects include fields of studies such as higher mathematics, arts, drama, music, writing, etc. Examining these subjects, one can see that they are highly viewpoint dependent. The common denominator of all of these subjects is the **viewpoint**!

"Mankind's greatest accomplishment is not the revolution of technology, it is the evolution of creativity."

—Anonymous

Photons in the Spectrum

Even though photons are used when dealing with the light frequency range, we can still refer to a quantum of energy in other frequency bands, such as the radio frequency (RF) or microwave bands, which we call respectively, an "RF Photon" and a "Microwave Photon."

Because photons of higher frequency carry greater energy, the *particle nature* of light becomes increasingly important as the frequency of radiation increases or as wavelength decreases.

The Figure on the next page shows the electromagnetic spectrum in terms of wavelength and frequency range and briefly delineates their modern applications as well as the scale of photon energy levels.

From this diagram we can see that photons in the lower infra-red range and below have energy levels below 1 electron volt (eV) and are usually considered waves, whereas photons with energy levels in the upper ultra-violet range and above have energy levels in the keV and are usually considered particles.

It is in the visible light range that we have a twilight zone, and a certain degree of confusion; an uncertainty of whether light acts as a particle or wave. This concept is treated in depth in the next article.

"There are two ways of spreading light: to be the candle or the mirror that reflects it."

—Edith Wharton (1862-1937)
American Novelist and Writer

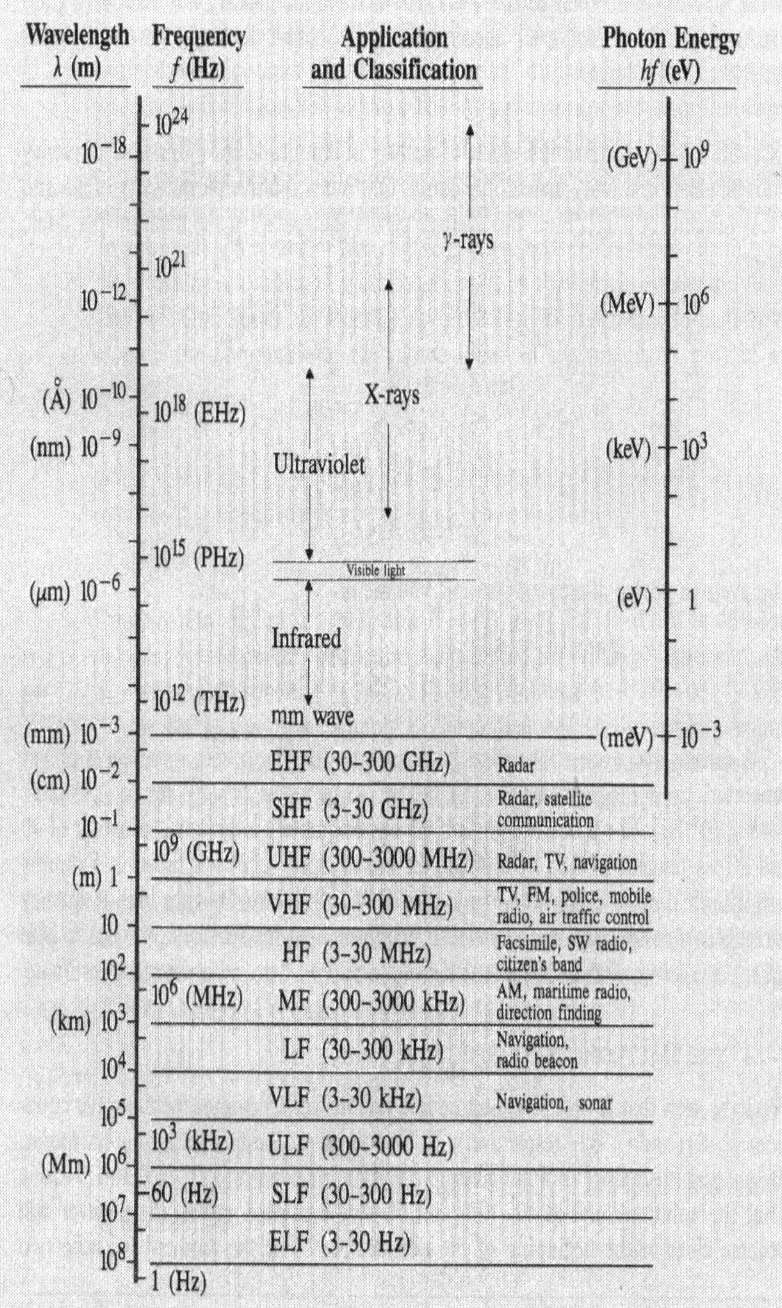

Electromagnetic Spectrum And Scale Of Photons

Particle or Wave?

Using the Heisenberg's uncertainty principle, we can conclude that for large scale phenomena existing on a classical physics level, the uncertainty principle does not pose any serious restrictions on the particle's location, velocity, energy, or time of observation.

On the other hand, for atomic and subatomic particles, the situation is indeterminate, because the uncertainty in position, velocity, energy, and time of observation is substantial to warrant a probabilistic study. Such a study is embodied in a subject commonly referred to as "quantum physics."

This brings us to one of the most common enigmas in the sciences today: "When is something considered to be a particle and when is it a wave?" To answer such a question, we need to break down the problem into its components.

In the first place, we can observe that a matter particle is a "condensed form of **energy.**" Secondly, a wave is considered to be "**energy** in motion." Thus, by a simple comparison we can see that both have "energy" as their common denominator. In other words, matter and energy are related at the source, since they are cut from the same cloth, figuratively speaking. This means that when a "particle" is in motion, it is actually "energy in motion," therefore naturally, we expect it to behave like a "wave." That expectation is expressed elegantly in the de Broglie's theory.

Upon further examination of the facts, we can observe that:

a) When we are considering a traveling "body" (such as a baseball), which is involved in large spaces, we are dealing with a definite particle. It is a particle with an exact location, because it is behaving on a classical level, which is deterministic and well known.

b) On the other hand, dealing with relatively small spaces that are on the order of atomic or subatomic levels, leads us to quantum physics, where the location of a particle (such as an electron) becomes indeterminate since its existence is uncertain and can only be expressed in terms of a probability. A particle's inexact location is now associated with a "wave packet," and thus, we have to describe it as a wave as shown below.

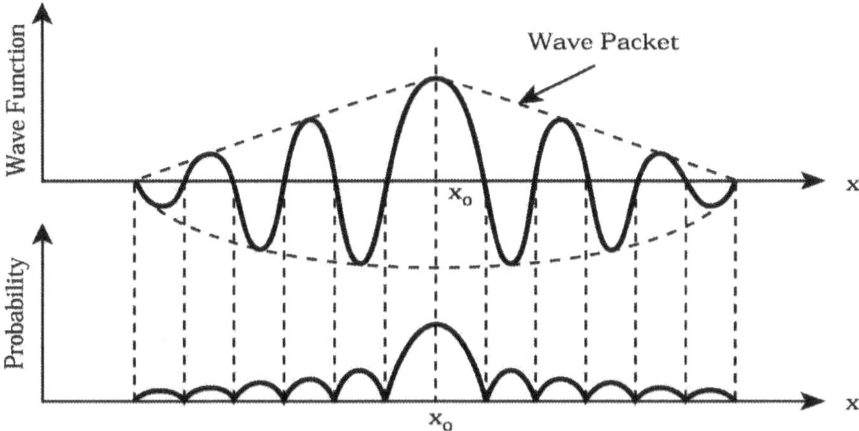

***A Particle's Wave Packet In Terms Of The Wave Function* (Upper), *And Probability* (Lower).**

From the above discussion, we can conclude that to understand the subject of particle-wave theory and answer the question of whether something is, "A particle or a wave," we need to ask a prior question, ***"What is the size of the created space in which the particle is placed?"*** *The answer will solve our dilemma. Thus, it is a question of size of the created space under observation that will provide the resolution to this riddle.*

"No problem can be solved from the same level of consciousness that created it."

—Albert Einstein (1879-1955)
German Born American Physicist, Nobel Prize in 1921

The Dual Theory of Light

Since wavelength (λ) of a wave and its frequency (f) have a reciprocal relationship ($f=c/\lambda$, c being the speed of light), the energy of a photon in electron volts (eV) can be calculated to be $1.24/\lambda(\mu m)$. Thus, we can see that shorter wavelengths have a higher quantized energy and behave more like a particle than a wave as shown below.

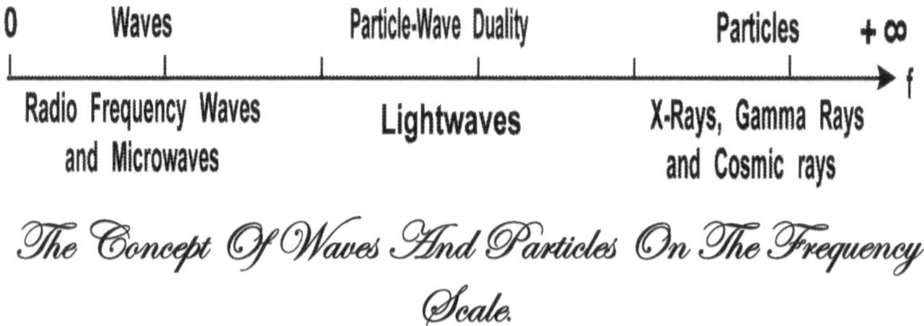

The Concept Of Waves And Particles On The Frequency Scale.

In other words, wave-like effects such as diffraction and interference become more difficult to discern as the wavelength becomes shorter. For example, X-rays and Gamma-rays ($f=3x10^{18}$ Hz) are very high frequency signals and operate at extremely short wavelengths, thus, almost always behave like a collection of particles (i.e., $E=1.24x10^{4}$ eV, a high energy-density particle); whereas radio waves and microwaves ($f=3x10^{10}$ Hz) are much lower in frequency and have much bigger wavelengths and almost always behave like waves (i.e., $E=1.24x10^{-4}$ eV, a low energy-density particle).

The frequency of light in the optical region ($f=3x10^{14}$ Hz) is such that both particle-like and wave-like behaviors occur (i.e., E=1.24 eV), which is a twilight zone of wave-particle existence.

The dual theory of light is based upon the observation and experiments about electron beams, being a stream of particles, which have wavelike properties and behave like light beams. This observation was confirmed by electron diffraction experiments, in which electron beams would diffract when a crystal structure was placed in their path.

The dual theory of light is obtained through years of observation and experimentation with light. It can be concisely stated as:

1) Light demonstrates a wave-like property in which it behaves like an electromagnetic wave as in classical physics. Furthermore, light also has a particle-like property, in which it behaves like a localized particle of matter as in quantum physics as shown below.

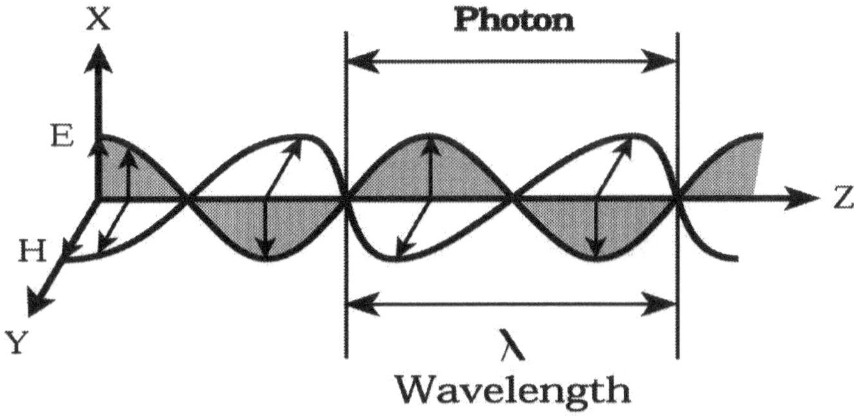

A Photon In A Sinusoidal Wave

2) A light beam of frequency (f) is composed of a stream of photons, particles possessing energy (E), which is proportional to its frequency multiplied by a constant called the Planck's constant ($h=6.62x10^{-34}$ Js).

3) The intensity of light decreases with increasing distance at any point in space, and is equal to a) The average power per unit area of the waves, or b) The density of the photons at that location is shown on the next page.

There is a strong similarity between light beams and electron beams in that they both travel rectilinearly, reflect and refract at an interface, and diffract in a similar manner.

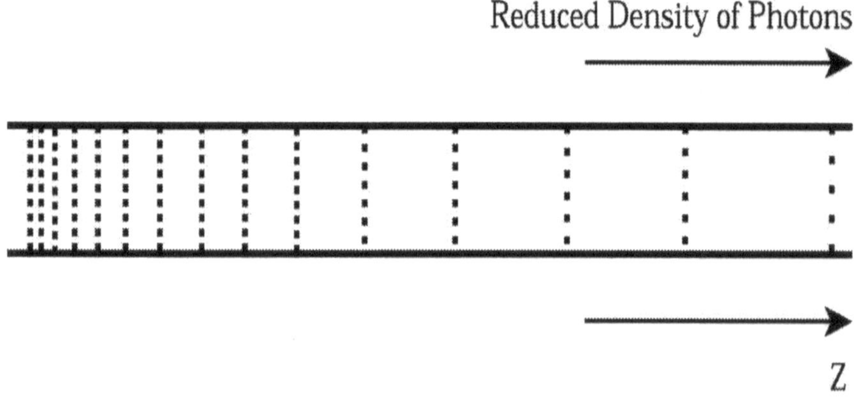

A Decrease Of Intensity Of Light.

(The closer the dashed lines, the higher the intensity)

Diffraction, in the case of a light wave, occurs due to a varying refractive index (such as a grating), which is analogous to the case of electron beams traveling through a varying force field (such as a crystal lattice). Mathematically speaking, both are analogous and equivalent.

Both of these phenomena can be described using classical wave theory. However, in the optical case, the refractive index changes over dimensions comparable to the light beam's wavelength, while in the case of an electron beam, the force field varies over a distance of the order of a de Broglie's wavelength (defined to be the wavelength of the wave associated with a moving particle of matter). In this case, we need to employ quantum physics (particularly Heisenberg's uncertainty principle) to explain the resulting probabilistic nature of the observed phenomena.

We can see that actual facts are quite contrary to the common belief of "mass of a photon being zero." This points out another pitfall in modern physics education where careless reporting of facts can cause havoc in one's study of the subject.

"In essence, science is a perpetual search for an intelligent and integrated comprehension of the world we live in."

—Cornelius Van Neil (1897-1985)
Dutch Educator and Pioneer of Microbiology

PART X

The Observable Application

Mass

The Source of Application Mass

"A man who as a physical being is always turned toward the outside, thinking that his happiness lies outside him, finally turns inward and discovers that the source is within him."

—Soren Kierkegaard (1813-1855)
Danish Philosopher and Theologian

It is interesting to note that the majority of scientists are often times fascinated and even astounded by the physical universe components and phenomena to a point that they forget these are in actual fact the application mass of a series of exact postulates that were made a very long time ago and have been carried through to their eventual reality. Example of physical universe components and phenomena include such things as sun, moon, earth, solar system, and galaxy on a large scale and electron, proton, neutron, atom, molecule and crystal lattice on a smaller scale.

At the present moment, we need to take a good look at the pyramid of knowledge and examine it rather carefully. This pyramid clearly shows that the highest point and the point of origin of any application mass is a series of postulates at the apex, which then and only then trickle down through axioms, natural laws, and other considerations to the final application mass of the postulates. This is a sweeping statement, which applies to any and all parts of life and existence.

Other than the written and recorded materials concerning the laws, theory and principles of operation of the subject (such as books, notes, etc.), the application mass is the main visible part of the postulate. The application mass is the entity, which is embodying the postulate made at the top of the pyramid and can be sensed and perceived with eyes or other senses, aided or unaided. It is subject to all the laws of the physical universe and evolves in its structure as time progresses.

The important point to grasp here is the fact that the application mass's existence is purely dependent on the existence of a postulate and not vice versa. Moreover, the fact that the application mass exists in a physical form clearly signifies the existence of the postulate.

"Don't confuse hypothesis and theory. The former is a possible explanation; the latter, the correct one. The establishment of theory is the very purpose of science."

—Martin H. Fischer (1879-1962)
German Physician, and Scientist

So we can make a sweeping observation that, "*Any visible and existing application mass signifies the existence of prior postulate(s), which produced it.*"

This conclusion is based upon the fact that an application mass is the byproduct of exact postulate(s) and not vice versa.

This is a significant conclusion, which should be used at all times in scientific study or research because application mass is junior in importance to the postulate that made it.

This concept, at first glance, appears to be easy to understand especially when the postulate is known and all the laws are derived, because then the design of the application mass is a relatively simple task.

However, in the practice of discovery and research, we encounter the reverse operation all the time. That is to say, we see a natural phenomenon, and we have to travel in the reverse direction (from base to apex of the pyramid) to reach the actual postulate that made that observed phenomena. That observed phenomena is really an application mass, which is backed up by an exact postulate. The discovery of the exact postulate would unravel the mystery and would bring about a total understanding of the observed phenomena. This is the discovery process as we know it today.

For example, a computer is an application mass because it is a visible entity. Of course it is a designed product made by man, therefore the designers have already discovered the digital logic postulates and have actually used them to derive a series of axioms, laws, and equations to aid them in the analysis or design process of the computer and many other similar digital devices. This is only the digital aspect of it. The series of postulates of "Boolean Logic" was originally proposed by a Greek philosopher named Aristotle over 2000 years ago and has been gradually put into a workable form over the last hundred years.

Then there is the implementation part of these postulates into a useful and exchangeable product, which requires an understanding of a series of physical universe postulates, a number of exact workable and implicit postulates of physics and engineering, a series of essential and auxiliary postulates of electricity along with the necessary technology to manufacture the product.

"Knowledge is a process of piling up facts; wisdom lies in their simplification."

—Martin H. Fischer (1879-1962)
German Physician and Scientist

The Significance of Application Mass

The existence of an application mass unravels the postulates that led to its current form as observed in present time.

Now through the use of inductive logic, we can generalize this principle to other natural application mass such as planets, stars, etc. This may be at first a relatively startling conclusion and a stretch of logic but a "natural" or "man-made" application mass follows the same pattern of development as delineated in the pyramid of knowledge. The natural application mass, which is not a direct byproduct of a science, is called the "*Generalized Application Mass*," in this work.

We see millions of stars and planets in our galaxy and get quite moved by the experience; however, we should recognize that these are all application masses of huge magnitude and follow the same pattern of creation. Yes, these are definitely some very gigantic masses occupying huge spaces, but the trick is to discover the postulate that went into making them and their mystery and awe would be solved instantly. Not who made the postulates, on which some may hang up (we leave that to religious philosophers to resolve), but rather what were the nature and exact sequence of the postulates that led us to have these minuscule things (such as atoms, etc.) or colossal entities (such as planets, etc.) that we refer to as application mass?

Physical sciences, being systematically a revolt against authoritative and religious dogma in 1500s, early on decided en masse to drop the question of, "Who did it?" and focused on a more productive question of, "How does it work?" or, "What is it?" This has been a very successful approach since it generated some of the most fantastic series of discoveries concerning the physical universe and opened the door wide open for the discovery of the rest of the galaxy. It allowed Man to look at things anew and investigate new phenomena without having to assume preconceived ideas concerning them.

In particular, scientists developed the "scientific methodology," which has enabled many to venture into new fields of study and become pioneers in those fields. This same methodology can now be leveled at the field of humanities and religion with tremendous success, to bring about a 180° turnaround in their dwindling spiral of success in explaining observed phenomena. This turnaround would certainly create a windfall of new information about these aforementioned subjects.

However, much to our disappointment, we find that the majority of religious philosophers, psychologists and sociologists are not scientists in the truest sense of the word. They have no formal training in the physical sciences, thus, are not aware of the scientific methodology or do not think that it is applicable to philosophy.

Therein lies the enormity of arbitrary solutions that have been generated through the ages in the process of solving man's personal or societal problems. They have no concourse with the uniqueness axiom, which is heavily used in physical sciences. Their methodology is not based on basic postulates or natural laws, and to that degree, they cannot and ordinarily do not find unique solutions to problems. Thus, their proposed theories and ideas are not workable on a general level, or even if workable, they are only useful in a very limited sphere of action.

"If most of us are ashamed of shabby clothes and shoddy furniture, let us be more ashamed of shabby ideas and shoddy philosophies. It would be a sad situation if the wrapper were better than the meat wrapped inside it."

—Albert Einstein (1879-1955)
German Born American Physicist, Nobel Prize in 1921

So in this work, we follow the tradition of the physical sciences and do not try to answer the question of "who?" or "why?" rather, a more tangible question of "what?" or "how?"

Our main line of thinking is not along the lines of, "Why does the viewpoint put forth a specific postulate?" or, "What are the composition and the main constituent components of a viewpoint?" rather, we are investigating the physical sciences along the lines of, "What are the series of postulates that led to the creation of a specific application mass?" or, "What are the sequential actions that a viewpoint needs to undertake to create an application mass?" The latter is our area of emphasis. This shift of focus (from " who or why" to "what and how") interwoven throughout the entire work, is as close to the basics as one can get without delving into a religious or philosophical discussion.

"The philosophy of one century is the common sense of the next."

—Henry W. Beecher (1813-1887)
American Liberal Congregational Minister

⊶ ✶ ❁ ✶ ⊶

Postulate and the Associated Application Mass

Having discussed the actuality of the physical universe as a generalized application mass, which is brought forth by a series of exact postulates, we can now use inductive logic to generalize this concept. This generalization leads to the concept of "Cause-Effect" as a pair in dichotomy.

The realization of "postulate" and "application mass" being a subset of "Cause-Effect" pair makes us aware of a much bigger picture concerning the general state of affairs, that is to say, the "universe of postulates" and the "universe of application mass" are opposite extremes of the simplicity scale, as delineated clearly on the pyramid of knowledge.

The "Cause-Effect" concept of "Postulate-application mass" can be understood more clearly if we delineate the properties of each and then draw our own conclusions.

A postulate, as the beginning point, of any creation can be considered to possess the following properties:

a) Zero space,
b) Zero mass,
c) Zero time dependency i.e. timeless,
d) Zero energy content,
e) An invisible quality.

An application mass (or a generalized application mass), as the end point of any creation can be considered to possess the following properties:

a) Finite space,
b) Finite mass,
c) Finite lifetime,

d) *Finite energy content,*
e) *A visible (or perceivable) quantity.*

Comparing the associated properties of postulates and application mass, brings about the inevitable conclusion that these two are opposite in their nature and must be treated as such in all aspects of our studies in sciences.

Furthermore, it can be observed that a postulate is the beginning of an imaginary road, which flows down through many conceptual layers and eventually ends in the application mass as the final destination.

From another vantage point we can see that the postulates are at the apex of the pyramid of knowledge whereas the application mass is at the base of the pyramid. From this vantage point the pair of "postulate and application mass" could also be further classified as an "essential and nonessential" pair, which clearly indicates the correct relationship between the two.

Thus, the inter-relation existing between a postulate and an application mass is inherent and inseparable. The important thing to understand here is that the flow (or the conceptual journey) from postulate to application mass is always one-directional, always from "cause to effect" (not vice versa).

The reverse direction, "Effect to Cause," is purely employed in research work for the discovery of some unknown principle or hidden mechanism of operation, to which inductive logic can be amply applied in order to develop a deeper insight into the operation of something or to extract a natural law, etc. In simple terms, we can call this process of going from effect to cause "detective work" or the "discovery process."

"Great ideas often receive violent opposition from mediocre minds."

—Albert Einstein (1879-1955)
German Born American Physicist, Nobel Prize in 1921

The Generalized Application Mass

Knowing the very basic postulates (i.e. the original and primary postulates) that formed the physical universe, we see that these postulates are embedded in every piece of matter, particularly commercial products and manufactured goods.

First let's enlarge and expand the concept of application mass, which has already been introduced earlier. The expanded concept of application mass, is "**Generalized Application Mass**," which we define as, "*Any created space, which contains created energies and created matter of any form, shape or size existing as a function of time.*"

In simple terms, generalized application mass is any matter and energy, condensed and packaged into an object form, which exists in a time-stream, from its inception to now. For example, in the field of electronics, a circuit board with all of its soldered components could be considered to be a good example of what we call the technical application mass, whereas the raw metallic ore found in nature is the generalized application mass. In a broader sense, a planet is another example of generalized application mass.

Now if we examine what a postulate is, we can see that, in actual fact, it is the viewpoint's considerations and decisions that govern the process of creation of application mass at the point of inception.

Thus, the essence of a postulate is a "consideration." It is the viewpoint (or the collective viewpoint) that says "so," and it will happen "so!" Therefore, through a cursory look and a simple examination of facts, we can conclude that considerations have seniority over the physical universe and its existing forms of created space, created matter, and energies. Thus, to fully understand the concept of generalized application mass, we need to examine a general subject called "considerations."

By **consideration,** we mean *mentally looking at something to create a thought about it, or specifically, directing the mind toward something in order to make a thought, a postulate, or a decision about its location, function, condition, etc.*

Therefore, we see that beyond the physical universe lies considerations. There are many examples that could be called forth to make this subject better understood. The examples could range from the large scale down to the very elementary level. The knowledge of Man as expressed in the physical sciences, is obtained by the observation of the mechanics and the rediscovery of the main postulates of that science as shown below.

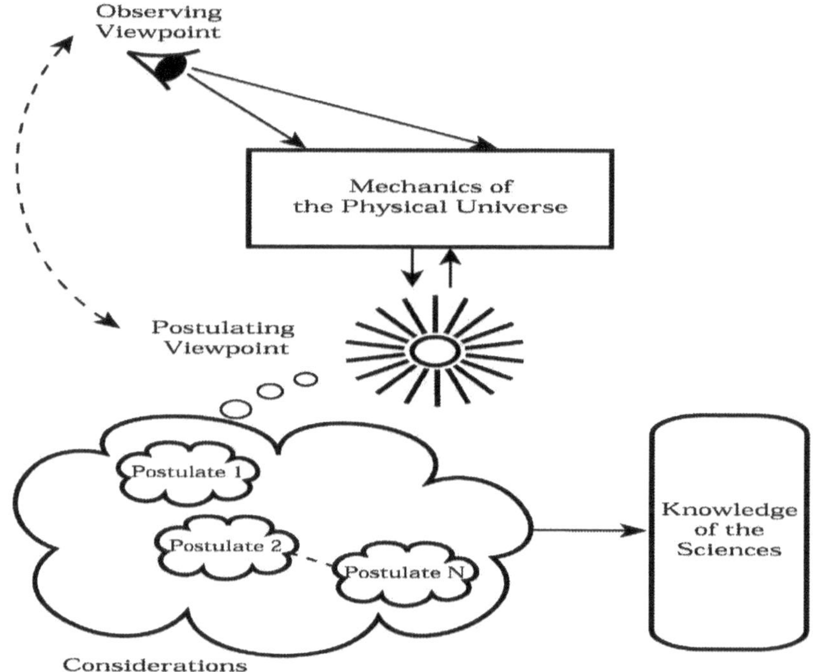

The Process Of Observing And Then Postulating, Leading To Knowledge.

Of course, one may argue that at this stage of galactic and planetary development, it is impossible to change them now, therefore they can not possibly be the results of postulates, and rather, they appear to be more like imposed facts.

This point is well taken but we should heed a basic factor here and that is once a postulate is made and put into effect by creating the corresponding application mass, one would have major difficulties in reversing it by “un-postulating” it. It will not work. It is really like getting a plot of land, drawing the architectural plans made, then pouring the foundation and erecting a ten-story building on that land. Then one day as the postulator of the building trying to change the building size by adding an eleventh story or try to erase the building by purely a thought process of “un-postulating” it. It would be an impossibility.

So is with the Generalized Application Masses (GAMs) called planets, stars, galaxies, etc. They will not cease to exist just because one “un-considers” them. They appear as imposing facts so long as one gives them life and considers them the all-important part of existence and ignores the fact that they are the derived byproducts and thus are extremely junior when they are compared to the postulates that created them.

The interesting point to note in all of this is that their current existence has followed an exact pattern of postulates, laid down in a precise sequence of action, which is unalterable at this stage of their development.

In other words, we are observing the material universe long after its inception point and are actually looking downstream, away from the apex and toward the base of the pyramid of knowledge.

Traveling down the pyramid of knowledge, one begets more and more complexity and less and less understanding. To find the answers to the riddles of our universe and achieve total understanding , one needs to go in the reverse direction, from effect to cause, from application mass to postulates, to find the truth.

"It is far better to grasp the Universe as it really is than to persist in delusion, however satisfying and reassuring."

— Dr. Carl Edward Sagan (1934-1996)

American Astronomer, Searching for Intelligent Life in the Cosmos

The Categories of Application Mass

"Your philosophy determines whether you will go for the disciplines or continue the errors."

—Jim Rohn (1930-)
American Motivational Speaker, Philosopher and Author

There are three categories of application mass that we may have indirectly alluded to in this work. However, we are at a point where we need to define each category specifically and exactly as discussed next. The Figure below shows the three categories of application mass.

"Science gives us knowledge, but only philosophy can give us wisdom."

—Will Durant (1885-1981)
American Writer and Historian

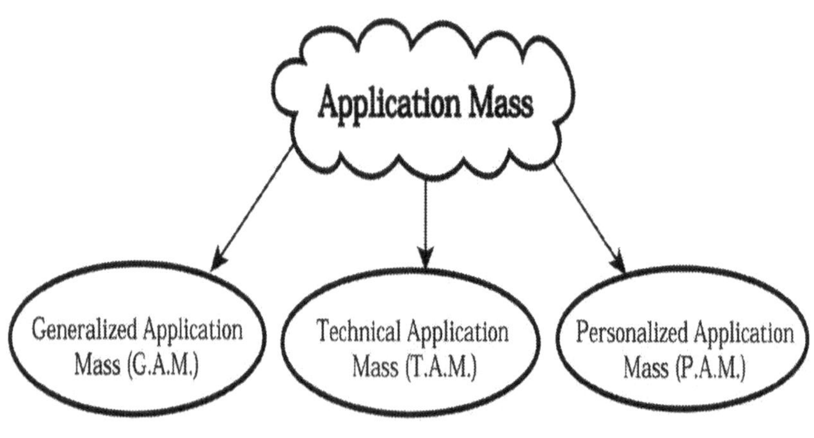

The Three Categories Of Application Mass.

Generalized Application Mass (G.A.M.)

Generalized Application Mass (G.A.M.) is the very general category of all masses (liquids, solids and gases) found in nature and existing in a raw and unprocessed form. Generalized Application Mass is obtained not through the postulates of a life force in a body form (commonly referred to as a human being), but one that is the result of "nature's postulates" put into practice, as though an invisible force or mind has executed them at some remote past and then put them on automatic to be faithfully carried out from then on.

The generalized concept of application mass includes the entire mechanical space containing all energies and all the existing gigantic masses of planets, stars, galaxies, which are not the direct byproduct of Man's sciences.

G.A.M. is that unthinking machinery, which we encounter everyday in our life, from the way atoms and molecules cohere together to form solids to the speed of revolution of earth around itself and the rate of rotation of earth around the sun, which defines our daily existence on this very forlorn planet. Examples of G.A.M. abound all around us and include things such as atoms, molecules, air, rocks, planets, shape and size of solar system, glaring rays from the sun, composition of galaxies, speed of our galaxy relative to others, so on and ad infinitum.

"To the mind that is still, the whole universe surrenders."

—Lao Tzu (600 -531 BC)
A Chinese Philosopher

Technical Application Mass (T.A.M.)

"Let no one delay the study of philosophy while young, nor weary of it when old."

—Epicurus (341-270 BC)

Greek Philosopher

Technical Application Mass (T.A.M.) is the category of man-made application mass that is produced directly as a result of application of a science using its scientific postulates, axioms, laws and other technical data.

It is a class of any physical entity derived from and related to the basic scientific postulates and technical facts. Examples include such things as a television set, a computer, an automobile, a power generator, a telephone system, a rocket, etc.

Occasionally we have used the word "application mass" to mean "technical application mass." Both may have been used interchangeably up to now, however, we have now broadened the concept of application mass as described above.

"It has become appallingly obvious that our technology has exceeded our humanity."

—Albert Einstein (1879-1955)

German Born American Physicist, Nobel Prize in 1921

⊶ ✶ ❁ ✶ ⊶

Personalized Application Mass (P.A.M.)

"Any religion or philosophy, which is not based on a respect for life is not a true religion or philosophy."

—Albert Schweitzer (1875-1965)
German Medical Missionary, Theologian, Musician and Philosopher
1952 Nobel Peace Prize

Personalized Application Mass (P.A.M.) is the category of application mass, which has been created and is based solely upon the viewpoint's own postulates and considerations. Examples of this category include such things as one's own customized possessions, any piece of artwork, jewelry, music, one's own body characteristics (such as hair-style, clothing, shape, etc.), a book's content and cover design, so on and so forth.

"Logic will get you from A to B. Imagination will take you everywhere."

—Albert Einstein (1879-1955)
German Born American Physicist, Nobel Prize in 1921

The Mechanics

Utilizing the concept of the three categories of application mass, we can see that regardless of its category, one still has to go through the very same process that is to say, starting with postulates at the top and moving down to the base of the pyramid of knowledge to reach the desired application mass.

The above three categories of application mass give us a panoramic view of the visible world, with which we find ourselves in direct contact everyday.

This panoramic view of application mass is equivalent to what philosophers normally refer to as, "**The Mechanics**," which is the mechanical aspects of life and existence. *Thus, we have connected the "physical sciences" to a much larger sphere of knowledge called "philosophy" through the use of the pyramid of knowledge, and actually have made them one of its many subsets.*

The totality of all types of application mass (i.e., G.A.M., T.A.M. and P.A.M.) is referred to as "mechanics," which is commonly used amongst philosophers and scientists to refer to anything not imbued with life force.

"Physical concepts are free creations of the human mind, and are not, however it may seem, uniquely determined by the external world."

—Albert Einstein (1879-1955)
German Born American Physicist, Nobel Prize in 1921

Since mechanics, in general, includes particles ranging in size from an electron to an object or a projectile (such as a missile, rockets, etc.) its study perforce requires the use of quantum mechanics at the microscopic scale and the laws of classical mechanics at the macroscopic level. For the study of super-sized particles such as planets or galaxies, we need to utilize the laws of astro-physics for proper interpretation of the observed data.

Any piece of matter can be in one category primarily but may have a combination of the other two categories, secondarily. For example, a sculpture is an artwork primarily (P.A.M.), but it additionally has T.A.M. (such as metal bars, cement, finishing oil, etc.) and also G.A.M. (such as atoms, raw stones, sand, etc.).

"Happiness is the meaning and the purpose of life, the whole aim and end of human existence."

—Aristotle (384 - 322 BC)
Greek Philosopher, Scientist and Physician

The Physical Body

The English dictionary defines the physical body as "*The whole physical structure and substance of a living entity in the form of a human being, animal or plant.*"

From the viewpoint of "Biology," the body could be considered to be a collection of trillion of cells functioning simultaneously together as an integrated unit. From the "Chemistry" viewpoint it is a chemical factory, which is in constant production with no down time. However, from the viewpoint of "Physics and Engineering," which is our primary focus in this work, the body is considered to be a rather complex engine equipped with an extremely elaborate *bandpass filter*.

The body, operating optimally at 98.6°F, is an extremely complex energy-conversion engine built and based upon the principles of physics, particularly "Thermodynamics."

Using the thermodynamics principles, we can see that the body works on the combustion (burning) principles in which the fuel (usually derived from other life forms) is combined slowly with oxygen to generate energy and heat. Since the body's building blocks are based upon the carbon atom (i.e., it is carbon-based), it could also be called a carbon-oxygen engine!

Functionally, since it has mass, it obeys all of the Physics' laws of motion such as Newton's laws of motion, etc. However, it is animated by the *life force* that occupies it.

"Begin to see yourself as a soul with a body rather than a body with a soul."

—Wayne Dyer (1940-)
American Motivational Speaker and Author

Thus, the body is subject to two sets of forces, a) Forces that arise from the life force to guide or move it in space (interior forces), and b) Forces that are exerted on it from the immediate physical environment surrounding the body (exterior forces).

From a totally exterior viewpoint, we see that the body is a detector. It detects waves such as light waves, sound waves, heat waves, motion waves, smell waves, etc. as well as other physical sensations and via the nervous system transmits these signals for processing to the brain, where it is picked up, analyzed and recorded on a mental level (both the physical signal and the accompanying conclusions) by the life force for actions and future utilizations.

"Life is the art of drawing without an eraser."

—John W. Gardner (1912-2002)
American Writer and Secretary of Health, Education and Welfare

The body as a detection instrument is a very *limited device* since its range of operation is extremely narrow; for example:

a) It detects only visible light at a peak wavelength of 0.555 microns (wavelengths in the range of 0.390-0.770 microns corresponding to photon energies of 1.8-3.1 electron volts). The visible light detectable by the eye corresponds to wave frequencies in the range of $5x10^{14}$ Hz to 10^{15} Hz.

Thus except for a tiny band in the electromagnetic spectrum, the remaining immense part of the spectrum, ranging from the very longest radio waves (very low frequencies of about tens of Hertz) to the shortest waves, such as Microwaves (10^9-$3x10^{11}$ Hz), Infra-red ($3x10^{11}$-$5x10^{14}$ Hz), Ultra-violet (10^{15}-10^{19} Hz), X-rays (10^{16}-10^{21} Hz), Gamma rays (10^{19}-10^{24} Hz), and cosmic waves (frequencies above 10^{24} Hz), is completely undetected by the human eye, or the body in general.

It is interesting to note that under high intensity radiation of short wavelengths (i.e. high frequencies of about 1 GHz and above), the body does detect such a high level of radiation on a very gross level and could be severely damaged depending on the intensity and the length of exposure. But this can not be called detection in the usual sense of the word, because it is accompanied by physical injury.

b) It detects sound in the frequency range of about 15 to 20,000 Hertz, and anything outside this range (such as ultrasonic or infrasonic sound) produces no response in a human body.

c) It can detect a limited temperature range of about 30°F to 100°F and outside this range without proper protection over a long period of time (such as heating or cooling), it disintegrates as a detector, i.e. dies off.

d) It cannot withstand and does not tolerate much force, without having an "alarming overload signal" commonly known as pain.

e) It has a top speed of about 10 m/s, which can only be maintained for a very short period of time.

f) The human eye cannot detect ordinarily a moving or changing scene beyond a rate of 25 frames per second, so on and so forth.

With all these limitations and shortcomings as a detector, there is a compensating factor, which saves the day! The human body has a supercomputer on board (called the mind), which is powered up by an immaterial energy source, completely separate and different than the physical energy that we see around us. The metaphysical energy, which is powering up the mind and our very physical existence, is manufactured and supplied by the "life force," which itself is an exterior force and not innate to the physical universe, and yet has a huge factor in its construction and orientation.

This supercomputer with its fantastic powers of observation, technical analysis and data evaluation, compensates for all of these deficiencies; it sees the invisible and the detects the unknowable by extrapolating beyond the range of input data and interpolating for the missing values in a range of input signals and comes up with practical solutions. Thus, it makes life on a physical plane possible!

"Our limitations and success will be based, most often, on our own expectations for ourselves. What the mind dwells upon, the body acts upon."

—Denis Waitley (1933-)
American Motivational Speaker and Author of Self-Help Books

On a higher note, we can see from a deeper echelon of knowledge, that the body could be understood to be the application mass (or more precisely, the Personalized Application Mass, P.A.M.) sitting at the bottom of the pyramid of knowledge, and its existence is purely dictated by the postulates that have gone into creating its existing form, characteristics and attributes whether positive or negative. The life force, as the driver of this sophisticated piece of machinery, assumes certain viewpoints, which mobilizes, guides, and directs it daily in specific directions (dictated primarily by purposes and secondarily by postulates at the top of the pyramid of knowledge), in order to enhance its survival potential on a physical plane of existence.

"Mind has no physical form and so it does not perish with the body; only the body perishes."

—Sri Sathya Sai Baba (1929-)
Indian Spiritual Leader

The Errors in Medicine

"Always laugh when you can. It is cheap medicine."

—Lord Byron (1788-1824)
English Romantic poet and satirist

It is interesting to note that there are two major errors in the field of medicine:

a) Confusing the body (the application mass) with the life force of the individual and giving it more seniority, which is an inverted look (i.e., an upside down view), and

b) Trying to approach the problems of the body purely on a physical basis, with a total neglect of the pyramid of knowledge in sciences.

Thus, the medical doctors may fail in their healing efforts to a greater or lesser degree, due to their failure to recognize the actual source of, "*The body problem*," which is equivalent to the application mass in our "Pyramid of Knowledge."

Utilizing the pyramid of knowledge in this arena, we can observe that the actual answers lie in the field of postulates. The field of Medicine, based upon its own limited perspective, is not able to embrace either, "*The New Science of Viewpoints*," or the ever dominant subject of "*Postulates*" at this moment!

"Reason and free inquiry are the only effectual agents against error."

—Thomas Jefferson (1762-1825)
American 3rd US President (1801-09),
Author of the Declaration of Independence.

PART XI

The Two Facets of Our Existence

The Two Interwoven Universes.

In actuality, we are dealing with a universe, which is two-faceted: a) One facet existing at a microscopic level, which is hidden and yet with extreme dominance underlies and actually makes up everything, and b) The second facet existing on a macroscopic scale, which brings about a level of reality with which we are too familiar and deal with on a daily basis as shown below.

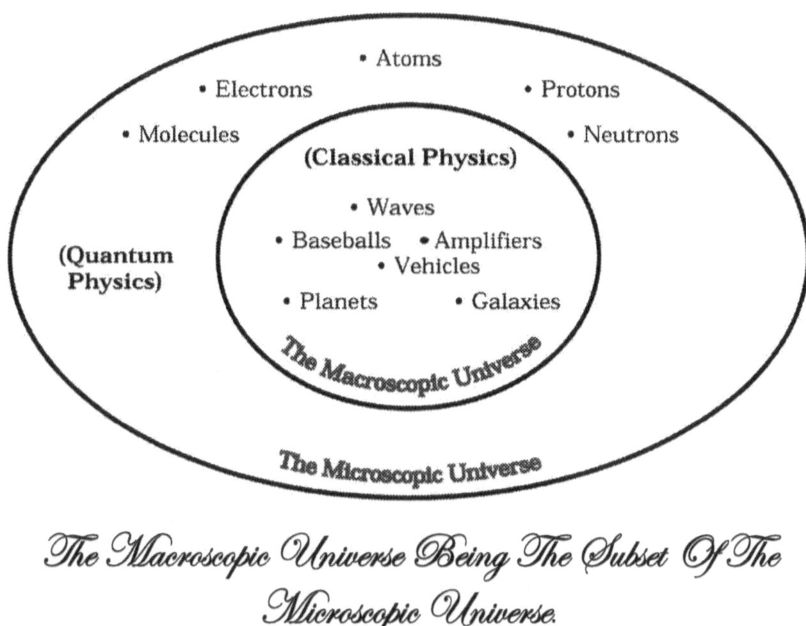

The Macroscopic Universe Being The Subset Of The Microscopic Universe.

The first facet is the actuality, whereas the second facet, dependent totally on the first, is an apparency. This is where the old maxim "change is the only constant in the physical universe" comes to life and becomes the ruling principle.

A deeper look into the physical universe reveals to us that it can be subdivided into two co-existing universes:

a) **Microscopic Universe** (Built upon ***Inherent Uncertainty*).**
b) **Macroscopic Universe** (Built upon ***Apparent Certainty***).

We need to define each at this juncture so that we may be able to, not only understand them better, but also make solid conclusions about the nature of our universe.

The Figure shown on the next page summarizes these two distinct and yet intertwined universes and clearly shows how the microscopic universe feeds the macroscopic universe.

Therefore, by looking across this arena and getting an inventory of our total knowledge and understandings, we realize that this is a rather confusing state of affairs when we speak of the physical universe. It is not an orderly universe where every motion and action in it is clearly marked, filed, and catalogued under an exact heading or even vaguely kept track of for future reference.

The beings in it make it seem orderly and organized to a very small degree but if the truth be told, this is a universe, which keeps itself hidden by constantly moving at a rapid rate and not telling us where it is going to. It asks one constantly to understand its confusions and find out its secrets.

The uncertainty inherent in the physical universe, which surrounds us at every aspect of our existence on a physical plane, has a dichotomy and that is the certainty in the viewpoint at the thought level.

This means that at the thought level, the viewpoint can assume a position of total certainty and understanding (the dual concept), which dissolves and replaces the uncertainty and confusion of the physical universe. In other words, certainty can be postulated into existence so as to make uncertainty junior in importance!

"All religions, arts and sciences are branches of the same tree. All these aspirations are directed toward ennobling man's life, lifting it from the sphere of mere physical existence and leading the individual towards freedom."

—Albert Einstein (1879-1955)
German Born American Physicist, Nobel Prize in 1921

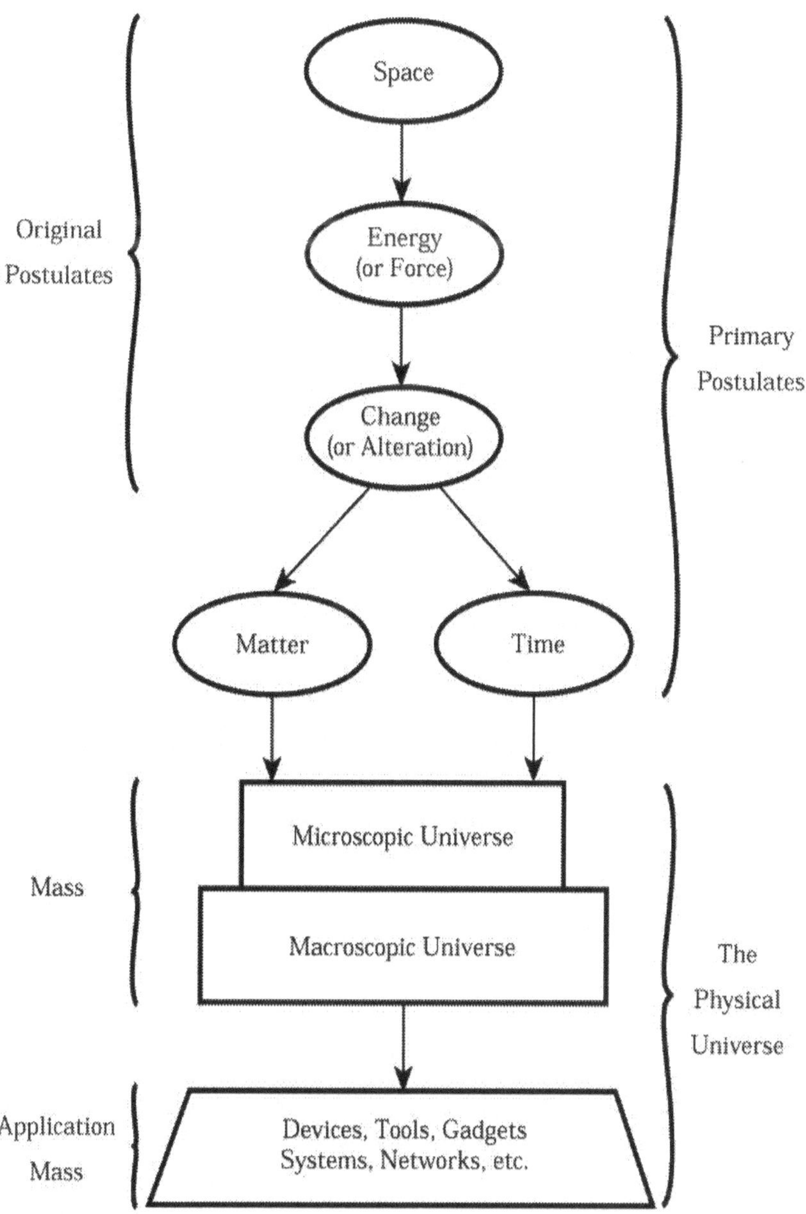

A Diagram Showing How The Two Universes Stack Up And Create the Application Mass.

The Microscopic Universe.

"If I have lost confidence in myself, I have the universe against me."

—Ralph Waldo Emerson (1803-1882)
American Essayist

The Microscopic Universe is the "actuality version" (also called the "underlying-theme version") of the physical universe, and consists of a series of light or heavy energy particles (photons, electrons, atoms, etc.) which are moving so fast that determination of their speed, energy or position with a high degree of certainty is impossible.

The microscopic universe actually feeds the entire macroscopic universe, thus, any and all of the observed motions and actions at the macroscopic level become glazed and laced with a layer of uncertainty (please see the "Uncertainty Principle" earlier discussed). This is where the axiom of "*absolutes are unachievable in the physical universe but only approachable*" takes its cue from.

"What men want is not knowledge, but certainty."

— Bertrand Russell (1872-1970)
English Logician and Philosopher

The Macroscopic Universe

THE MACROSCOPIC UNIVERSE is the "apparency-version" (also called the "visible version") of the physical universe and consists of a series of many particles grouped together where each group of particles is formed into one macro-particle and acts as one coherent unit. These macro-particles are usually referred to as objects or bodies (such as baseballs, cars, airplanes, planets, galaxies, etc). The objects at this level of observation follow certain and definite laws of action or motion, which behaviorally are very predictable and mathematically, very precise.

At the macro-particle level, we have an apparency, where it seems that the macroscopic version of the physical universe is understandable and appears to bring forth a high level of certainty and predictability with it.

However, since the microscopic version of the physical universe is the underlying theme of all things, therefore it is the most dominant and overriding factor in all of this.

On a lighter side and metaphorically speaking, we could think of the microscopic universe as raining uncertainty on the parade (a parade of predictability and certainty of motion and action) that macroscopic universe presents to us everyday for viewing. This would surely classify as raining on someone's parade with exclamation point!

"My religion consists of a humble admiration of the illimitable superior spirit who reveals himself in the slight details we are able to perceive with our frail and feeble mind."

—Albert Einstein (1879-1955)

German Born American Physicist, Nobel Prize in 1921

The Two Fields of Physics

"Science is a way of thinking much more than it is a body of knowledge."

—Dr. Carl Edward Sagan (1934-1996)
American Astronomer, Searching for Intelligent Life in the Cosmos

Physics is the science dealing with energy and matter and the relationship between them, such as their interactions, motion, forces, properties and changes in energy states excluding biological and chemical changes. Physics is divided into two fields:

a) **Classical Physics**, which itself is further subdivided into electricity, magnetism, optics, acoustics, mechanics, and heat; and
b) **Quantum Physics**, which is subdivided into atomic physics, nuclear physics, and solid-state physics.

'Physics is, hopefully, simple. Physicists are not."

—Edward Teller (1908-)
Hungarian Born American Nuclear Physicist

Physics is not focused upon any specialized form of matter, as is chemistry (which focuses upon the properties, composition, structure, and interactions of matter and the energy changes that accompany these interactions), or as is biology (which focuses only upon the physical component of life, which is not just matter but matter imbued with life force).

Matter, in physics, is anything which has mass and occupies space, and for a very long time, it was considered to be distinctly separate and different from energy. However, with the increased knowledge of the nature and through several important discoveries, matter has been shown to be fundamentally a condensed form of electrical energy.

In fact, through the Einstein's formula ($E=mc^2$), we can see that the mass of a material particle (m) and its energy content (E) are equivalent through a proportionality constant (c^2).

Therefore, we can conclude that higher physics and chemistry have become interrelated, and have actually taken us one step closer to a much bigger concept and that is "The universal nature of energy."

"The object of pure physics is the unfolding of the laws of the intelligible world; the object of pure mathematics that of unfolding the laws of human intelligence."

—James Joseph Sylvester (1814-1897)
English Mathematician, Developed Number and Matrix Theory

Quantum Mechanics

"The whole of science is nothing more than a refinement of everyday thinking."

—Albert Einstein (1879-1955)
German Born American Physicist, Nobel Prize in 1921

In the early part of the twentieth century (circa 1920), it became necessary to develop a new theory to describe phenomena on the atomic level. A number of observations concerning atoms and electrons and their behavior had indicated that they did not obey the governing laws of classical mechanics or physics.

The experiments that led to the development of quantum mechanics were concerned with the nature of **light** and the relation of its energy to the energies of **electrons** within atoms. These experiments provided enough indirect evidence on the nature of energy on an atomic scale, to warrant the development of a new theory in order to provide coherent answers to the mysteries of the physical universe at such microscopic levels of created space.

Thus, at that point in time, a search had begun to develop a new theory. Many brilliant minds such as Einstein and Planck and others undertook this challenge and brought the matter off and thus, filled the void for this much needed theory. They saw that it had become imperative that a new kind of mechanics be put into use; one that would embrace and explain the behavior of particles on a microscopic scale. Thus, quantum mechanics (also called quantum physics) was born in response to this need.

Through the years, the amount of information in quantum mechanics has grown to a point that it stands now as tall as classical mechanics and, in combination, provides a valid description of all natural phenomena.

Since the theory of quantum mechanics is mostly expressed in complicated mathematical expressions and is somewhat separated from the reality and observation of our day-to-day existence, many students have difficulty understanding or relating to it. Most find it more comfortable to deal with the classical mechanics since it is much easier to observe, experience, and understand its many aspects directly and on a first-hand basis.

The effects of atoms and electrons can only be observed through indirect measurement and by instrument read-out, thus our own first-hand observation and experience is impaired due to such a small scale of observation.

Therefore, the need to depend on the abstract concepts of quantum mechanics to predict the experimental results on an atomic scale rather than to attempt to use classical mechanics is even more important. This is such an important point to grasp that cannot be overemphasized. In other words, we can not force our understanding of classical explanations into the nonclassical phenomena of atomic and subatomic particles, and still expect to come away with a correct answer.

Therefore, quantum mechanics has filled a huge gap in our knowledge about the physical universe's finer particles, which are invisible to the naked eye, and yet influence and underlie many aspects of our observed natural phenomena on a daily basis.

"You and I are essentially infinite choice-makers. In every moment of our existence, we are in that field of all possibilities where we have access to infinity of choices."

—Deepak Chopra (1947-)
Indian Physician and Author

Classification of Motion

"Our nature consist in motion; complete rest is death."

—Blaise Pascal (1623-1662)
French Mathematician, Philosopher and Physicist

The concept of motion can generally be classified into two categories, which forms an important part of scientific knowledge as follows:

I. QUANTUM-PHYSICS TYPE MOTION (ALSO CALLED MICROSCOPIC OR SMALL MOTION)—This type of motion deals with motion of particles on: a) atomic, and b) subatomic scale, which are governed by the laws of quantum physics. This type of motion could also be called "small motion." For example, let us consider an atom, where it is composed of many electrons and a nucleus. Each electron has a spinning motion about itself while also moving in an orbit around the nucleus. This particular sub-division of kinetic energy leads to study of atomic and subatomic particles, a subject called quantum physics as shown on the next page.

II. CLASSICAL-PHYSICS TYPE MOTION (ALSO CALLED MACROSCOPIC OR LARGE MOTION) - This type of motion deals with:

a) Motion of matter particles, which are much grater in size than an atom and at speeds, which are much slower than the speed of light, and
b) Propagating energy fields in large spaces (large relative to their wavelength), which can be expressed by continuous functions of time and position

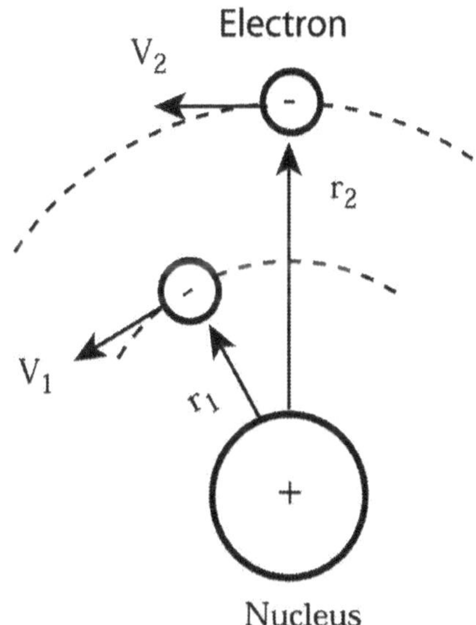

A Nucleus And Orbiting Electrons Comprising An Atom.

This type of motion could also be called "large motion." Examples for this category of motion abound in our environment, for instance, an example of (a) would be a moving baseball and an example of (b) would be wave propagation in a transmission line such as in a TV cable.

"If anybody says he can think about quantum physics without getting giddy, that only shows he has not understood the first thing about them."

—Niels Bohr (1885-1962)
Danish Physicist, First to Apply the Quantum Theory,
Nobel Prize for Physics in 1922

Types of Large Motion

"The aim of every artist is to arrest motion, which is life, by artificial means and hold it fixed so that a hundred years later, when a stranger looks at it, it moves again since it is life."

—William Faulkner (1897-1962)
American Short-Story Writer and Novelist, Nobel Prize for Literature in 1949

The large motion which is under the heading of classical-physics type motion breaks down into three classes of motion:

a) A FLOW,

b) MULTIPLE-FLOWS,

c) A STANDING WAVE.

These concepts are explored in the next few pages.

"There must be a positive and negative in everything in the universe in order to complete a circuit or circle, without which there would be no activity, no motion."

—John McDonald (1916-1986)
American Writer

A Flow

"A republican government is slow to move, yet when once in motion, its momentum becomes irresistible"

—Thomas Jefferson (1762-1825)
American 3rd US President (1801-09),
Author of the Declaration of Independence

A FLOW is a transfer of energy or matter particles, objects or waves from point A to point B. This is a very simple concept dealing with a particle moving in a definite direction. The particles (such as electrons, atoms, etc.) always flow from *higher to lower potential energy levels*, where the difference between the two levels is converted to the kinetic energy in the moving particles.

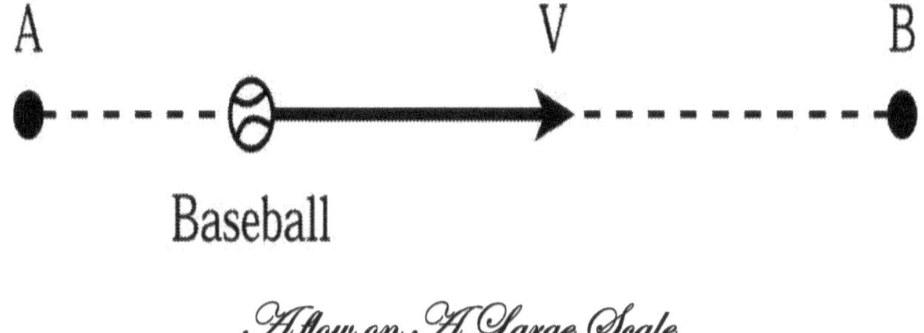

A flow on A Large Scale

Flow examples are water flow in a river, wave propagation in a TV cable, electricity flow in a wire, etc.

"Never confuse motion with action."

— Benjamin Franklin (1706-1790)
American Statesman, Scientist, Philosopher, Printer, Writer and Inventor

Multiple Flows

MULTIPLE-FLOWS OR A FLOW DIVERGENCE (ALSO CALLED A DISPERSAL OF ENERGY) is the generalization of the concept of a flow and is defined to be many outflows (or inflows) of particles from a common point. In other words, this term embraces the concept of "multiple flow" having a common origin.

This is very similar in concept but not synonymous with what is commonly referred to as, "Divergence of a vector quantity," in physics, which is a vector operation measuring the net outflux (or the influx) of a vector quantity from a point in space.

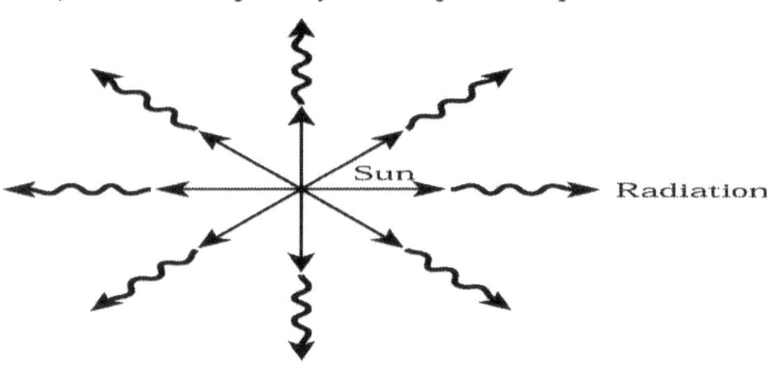

A Flow Divergence (Or Dispersal Of Energy)

Divergence examples are water sprinklers for the lawn shooting water in a circle; The Sun radiating out spherically on a constant basis; Fireworks used in celebratory occasions, particularly on the Fourth of July in U.S., etc.

"The quality of the imagination is to flow and not to freeze."

—Ralph Waldo Emerson (1803-1882)
American Essayist

Standing Waves

"Like as the waves make towards the pebbled shore, so do our minutes, hasten to their end."

—William Shakespeare (1564 - 1616)
English Dramatist, Playwright and Poet

A STANDING WAVE (ALSO CALLED A RIDGE OF ENERGY) is caused by two energy flows, impinging against one another, with comparable characteristics to cause a suspension of energy particles in space, enduring with durations longer than the flows themselves.

It is important to note that in order to form a standing wave, we must have the characteristics of the two flows to be comparable in all aspects. Such characteristics include type, frequency, amplitude, etc.

"Science is nothing but perception."

—Plato (428-348 BC)
Ancient Greek Philosopher

In other words different types of signals, with different amplitudes and frequency do not form standing waves. An example of something that would not form a standing wave would be a water flow running against an air flow, light waves colliding with microwaves, etc.

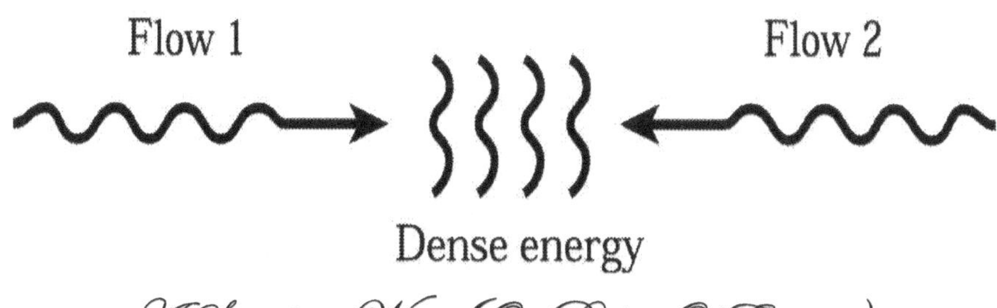

A Standing Wave (Or Ridge Of Energy).

Standing Wave examples are any piece of matter in our environment is a standing wave on a microscopic level, two cars colliding, two waves colliding on a pond, etc.

The main difference between a standing wave and the other two types of motion is that a standing wave has a "location" whereas a flow has a "direction" and a divergence has "many directions." The frequency associated with a standing wave corresponds to the frequency of one of the flows before the impingement.

Classical-physics type of motion and its associated kinetic energy is an area of primary focus in the majority of physics texts, without a mention of its counterpart: Quantum physics.

"To raise new questions, new possibilities, to regard old problems from a new angle, requires creative imagination and marks real advance in science."

—Albert Einstein (1879-1955)
German Born American Physicist, Nobel Prize in 1921

Nano-Technology

"Great acts are made up of small deeds."

—Lao Tzu (600 -531 BC)
A Chinese Philosopher

One of the direct applications of the microscopic universe is the newly emerging field of "nano-technology," where "nano-"means one billionth (10^{-9}) of anything. Nano-technology deals with the microscopic universe and, therefore, could be considered to be a subset of **quantum mechanics**. It is the science of working at atomic dimensions to engineer materials and machines out of individual molecules, and is based upon the same set of postulates and axioms that we have discussed so far.

Nano-technology utilizes directly the microscopic universe to create macroscopic effects. Examples are many, but a few would include the following:

a) A Battery for laptop computers that will last for weeks (not hours) without recharging.
b) A microscope that works at nano-level displaying spaces five atoms apart on a semiconductor wafer (such as Silicon, Gallium Arsenide, etc.) to help etch the imprinted circuits on a microchip more finely.
c) A whole bio-laboratory on a chip to constantly monitor water supplies for minute amounts of impurities in order to certify the quality of the drinking water.

The landmark discovery that sparked and opened the door to nano-technology research, occurred only a few years ago when carbon nano-tube materials, thinner than human hair, were developed. These carbon nano-tubes, when integrated into macroscopic objects, created such strong and yet light materials that they could replace commonly used metals (such as steel) in many applications.

These new nano-tube carbon materials could have wide applications in such things as automotive body parts (such as fenders, bumpers, etc.), bulletproof fabric in a soldier's military uniform, so on and so forth. Furthermore, these materials have a very low electrical resistivity, making them suitable for transmission lines and communication circuits, which can be built right into the military uniforms.

The next level of development will point in the direction of making molecules "replicate themselves," just like cells do in a body. This concept has a much higher ramification, since we could have machines that repair and build themselves back up again after performance failure due to physical malfunction. This could be a great application in space vehicular technology, where sending a repair crew is costly and not frequently available.

The idea of "molecule replication" could give us a higher survival rate for machinery and equipment and thus, bring about a whole new level of immortality for the technical application mass (T.A.M.) , in general. There could be light bulbs that never burn out, cars that run for years on a single charge, and sport supplies and equipment (such as tennis balls, etc.) that last practically forever without going flat.

The concept of Nano-technology, if brought forth to fruition, is likely to affect every aspect of our lives in the future since the sheer number of many applications, that could possibly use this technology, is limitless!

"Anyone who doesn't take truth seriously in small matters cannot be trusted in large ones either."

—Albert Einstein (1879-1955)
German Born American Physicist, Nobel Prize in 1921

The Measurement Systems

The physical universe, as we observe a portion of it today, is actually based on three inexorable original postulates, which have led to four primary components of the physical universe.

Each primary component of the physical universe, through millennia of observation, has gathered to itself systems of measurement, which provide the inhabitants of this planet, accurate information about the magnitude or size of these components, so that correct estimation of effort can be engaged upon, in order to perform a desired task as shown on the next page.

From this figure we can see that the important measurement systems useful to scientists are:

1) Space Measurement systems
2) Energy Measurement systems
3) Time Measurement systems
4) Mass Measurement systems
    a) Micro-scale: Atomic or molecular level.
    b) Macro-scale: Object or planetary level.

These four levels of measurement are employed everyday on this planet and on a constant basis, to produce meaningful numbers, which can be used effectively to estimate the effort needed to perform many complicated tasks properly and without accidents. For example, the exact propelling force needed to lift an aircraft off the ground, the exact current needed to power up our computer, the exact moment of discharge in a spark plug to ignite the gas in the car engine for a smooth ride, etc., are all vital values that need to be known exactly by accurate measurements.

"Any measurement must take into account the position of the observer. There is no such thing as measurement absolute, there is only measurement relative."

—Jeanette Winterson (1961-)
English Writer

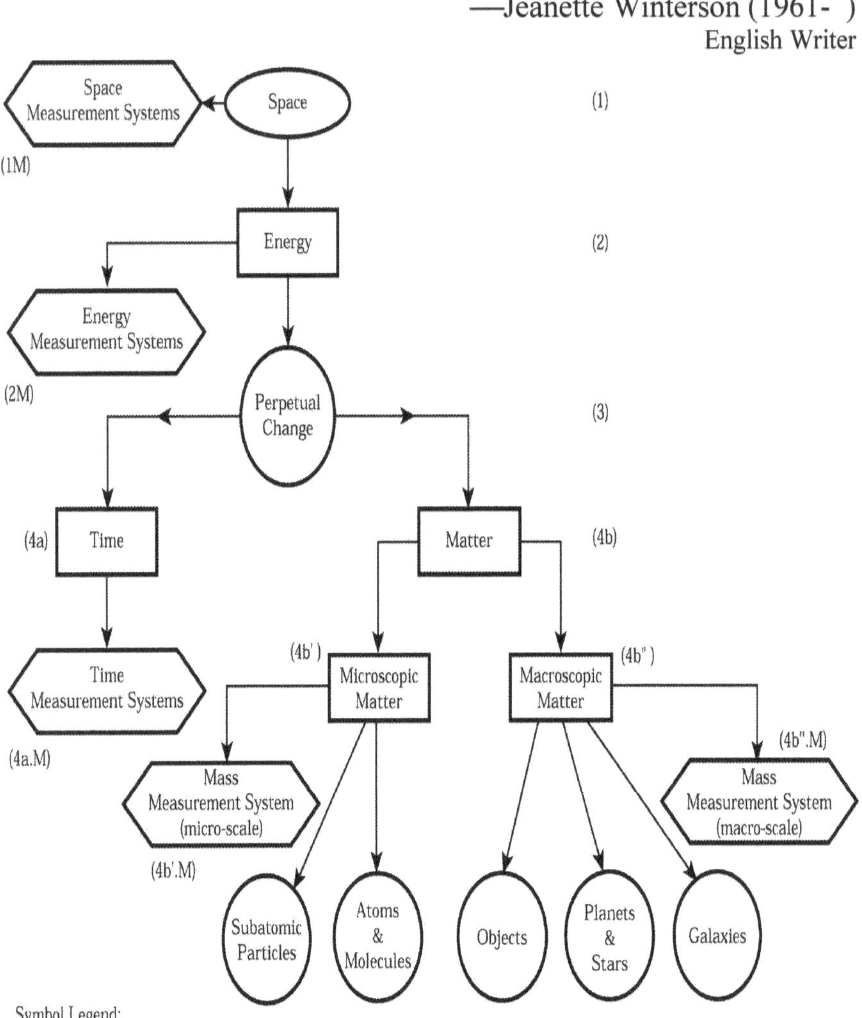

Levels Of Measurement Systems Employed To Measure The Basic Quantities In Our Universe.

PART XII

The Crowning Secrets

The Physical Universe in a Nutshell

The consecutive series of postulates that have gone into the construction of our physical universe is diagrammed on the next page.

These postulates were thought of consecutively at the blue-print stage but at the implementation stage, have been put forth and implemented simultaneously.

The two main players of this universe are "energy" and "matter," which are in constant interaction. Their interplay actually causes the so called "mechanical time" as a byproduct of the interaction of the two.

Earth is a small planet in the solar system, a tiny dot in the Milky Way galaxy, which itself is but a small subset of the vast expanse of what we commonly refer to as the "physical universe." Long before mankind appeared on earth (about 35-100 thousand years ago), the panoramic physical universe had come into existence and was set up as a habitable planet for the biological organisms amongst which homo sapiens, as a species, is but a subset.

"A human being is part of a whole, called by us the Universe, a part limited in time and space. He experiences himself, his thoughts and feelings, as something separated from the rest, a kind of optical delusion of his consciousness. This delusion is a kind of prison for us, restricting us to our personal desires and to affection for a few persons nearest us. Our task must be to free ourselves from this prison by widening our circles of compassion to embrace all living creatures and the whole of nature in its beauty."

—Albert Einstein (1879-1955)
German Born American Physicist, Nobel Prize in 1921

The diagram below shows a summary of the postulates that have gone into the construction of the physical universe.

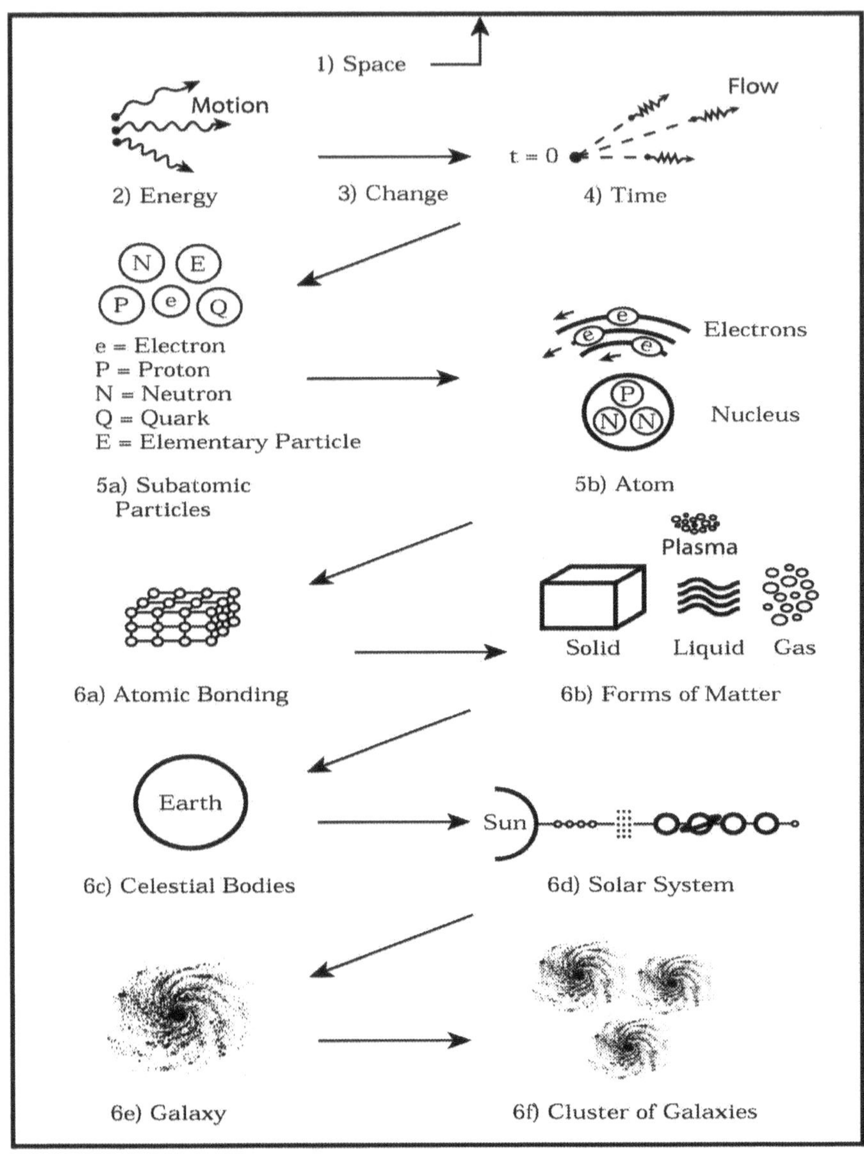

A Bird's-Eye View Of The Postulates That Make Up Our Panoramic Physical Universe.

The Universe Ahead

"Just remember - when you think all is lost, the future remains."

Dr. Robert H. Goddard (1882-1945)
American Rocket Engineer

A number of points have gradually emerged as the essential factors in making a universe. One of these essential factors, besides a viewpoint, is postulate.

Therefore as one looks at any new universe, the first visible thing that one encounters is the application mass of that universe. These are the created spaces, wherein all of the created masses and energies are placed. Of course, the one thing mostly missed in this process is the postulator and the series of postulates that went into making that created mass or created energy.

"A mind at peace, a mind centered and not focused on harming others, is stronger than any physical force in the universe."

— Wayne Dyer (1940-)
American Motivational Speaker and Author

Therefore, if one wishes to delve into what lies beyond the physical universe or for that matter how to make another universe similar to or completely different from an existing universe under study, one needs to look no further than the following points:

a) One needs to understand the basic postulates or considerations that have been utilized in the construction of any universe under study (such as the physical universe, one's own universe, etc.),

b) One needs to create or make new postulates or considerations such that they either generalize or encompass the old ones, which then allow a basic change to occur in the universe one is studying, and

c) *Finally, based on the new postulates, one derives all of the related axioms, laws and design techniques and builds a whole gamut of new application mass to replace the old ones. This process would surely bring about a whole new universe or field of study and can be called "creating a new universe."*

If one is in constant agreement with a universe (its postulates and laws) and keeps remaining in that universe and yet intends to change that universe, one is really fighting one's own earlier postulates and would not be able to change that universe. This is because any new postulates will be fighting the earlier ones and thus, will create a problem or a standing wave of two postulates, where the earlier ones will be dominating and will create an apparency or an apparent status quo of "no change." To remedy this situation, one needs to locate the actual and fundamental postulates of that universe and by disagreeing with them, eradicate them and then create new ones (usually of comparable magnitudes) to replace them.

For example, over 2000 years ago the Greek philosopher Aristotle postulated the "Digital Universe," which is based upon the two-valued logic system (the true or false system) consisting of three main postulates. With this, he ushered in and thus, established a whole new field of reasoning and study, which we now call "digital logic." Over a period of two thousand years we have gone through quite a bit of evolution and gradually but finally have reached its associated application mass in terms of numerous digital devices, circuits, components, computers, and systems.

"Once you make a decision, the universe conspires to make it happen."

—Ralph Waldo Emerson (1803-1882)
American Essayist

Beyond Postulates: Purposes

Let us briefly take stock of our knowledge base and evaluate what we know so far about the physical universe. Such an action can best be described by a simple tabulation of our knowledge base:

a) *First there is a viewpoint,*
b) *Then a series of postulates are put forth,*
c) *Next, a number of laws are derived through unbiased observation and study. These regulate the interaction between the postulates and set down the required ground rules to operate within that sphere of activity, and*
d) *Finally, we have various shapes and forms of application mass, which are the visible byproducts of the postulates, and collectively we refer to as a "universe."*

Now, if we look carefully between the (a) and (b) levels we may sense a tiny gap, which we have put aside up to now and have not cared to explain it, and that is, "Why does a viewpoint make such a set of postulates and no other?" After all, at the blue-print stage of any universe, a viewpoint is free to make any postulates, but, "Why this set and no other?"

The above posed question can be adequately answered by a careful examination of the subject and the realization that postulates are made for a "reason" and that reason is all under the heading of "purposes."

Therefore we can conclude that in general, a series of postulates are made solely based upon a series of purposes. The purposes dictate the extent and shape of the postulates.

"Every moment of your life is infinitely creative and the universe is endlessly bountiful. Just put forth a clear enough request, and everything your heart desires must come to you."

—Mahatma Gandhi (1869-1948)
Indian Philosopher

We will not go so far as to surmise what were the purpose(s) of the viewpoint(s) in creating the four primary postulates, which led to the creation of the physical universe. The answer to such a question would put us into a speculative level of operation and that would be beyond the scope of this work as shown below.

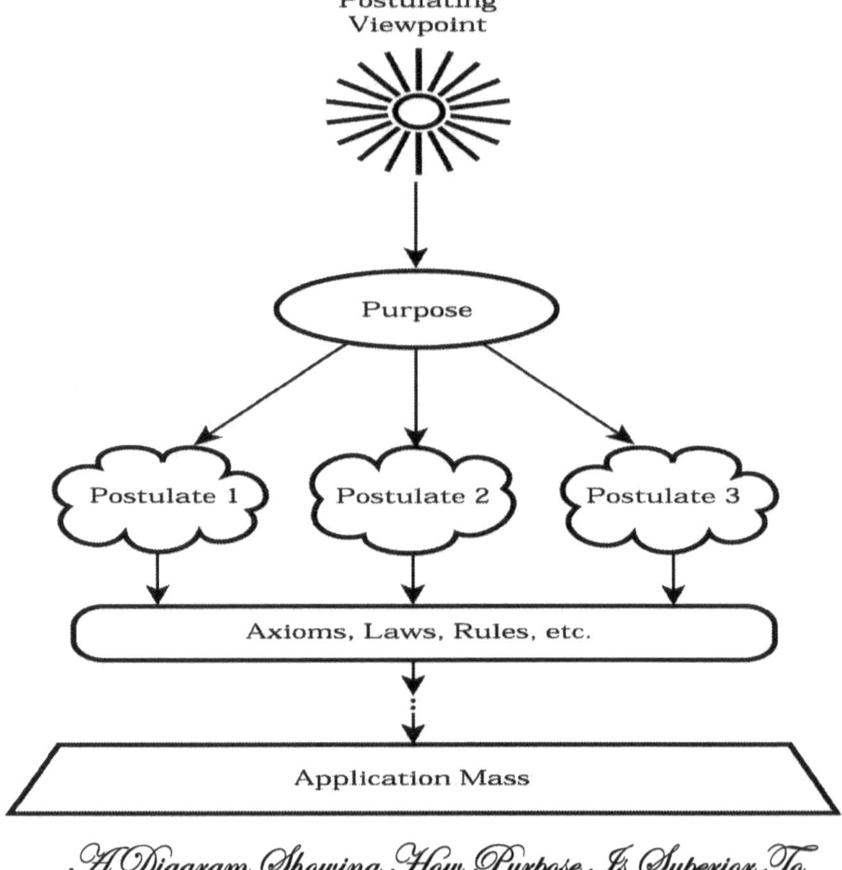

A Diagram Showing How Purpose Is Superior To Postulates.

"Be above it! Make the world serve your purpose, but do not serve it."

—Johann W. von Goethe (1749-1832)
German Poet and Novelist

Resorting to our earlier example of building a house by first dreaming about it (i.e., postulating it), designing and then building it, we can see that the "purpose" of the postulator or the viewpoint can be sensed and is actually written all over the house. For example, the postulator wanted a four-bedroom house because he has a purpose in raising three kids. He has a purpose in aesthetics and likes the view of the ocean, so he orients the dining room overlooking the ocean. He has a purpose in growing vegetables and flowers, so he postulates a flower patch near the house and a vegetable garden in the back yard. The list goes on.

Therefore, we can sum up our thoughts by realizing that addition of purposes to a series of postulates (made by a viewpoint) makes the set of creation on a mental level complete and the path wide open for anyone to attempt to bring a universe of one's own into existence. Now one's newly postulated universe may not be as vast or as dominant as the physical universe but it surely is a universe, and surely one knows it is his because he built it from scratch, all based on the solid principles presented in this book.

"There is one quality which one must possess to win, and that is definiteness of purpose, the knowledge of what one wants, and a burning desire to possess it."

—Napoleon Hill (1883-1970)
American Author

Recovery of Lost Postulates

The fact that we have to recover a postulate in the discovery process indicates that a physical postulate is susceptible to getting lost and being put out of sight as time marches forward. This brings about an interesting and terribly unfortunate state of affairs, where not only in our society amongst the average individuals but more importantly in the scientific communities across the land and amongst scientists, no one seems to pay any attention to the fundamental postulates, which form the woof and warp of our current existence.

For example, the extant scientific literature most often discusses the laws, rules, techniques, etc. of a certain observed phenomenon and fails to present what is occurring on a thought level behind the scene. The technical presentations in classrooms, at the level of higher institutes of learning all across the globe, are rather robotic. Most educators, scientists and engineers use mathematical manipulation of symbols to supplant for understanding, with little or no discussion at the postulate level!

"Success is going from failure to failure without losing enthusiasm."

—Winston Churchill (1874-1965)

British Orator, Author and Prime Minister During World War II

The physical universe happens to be a "consequence-happy" universe, which means that for every action one may undertake in it, one should be prepared to receive a consequent reaction, which as a result has turned it into a dangerous and harsh universe. It does not care to understand but only wants to be understood. For example a cup falls off of a table; it receives instantly a reaction and shatters without any thought from the physical universe. To put it mildly, it is an "unthinking universe of force." It has no concourse with reason or logic.

The beings in it make it look less dangerous by doing its thinking for it. They impede and dampen its direct and harsh reactions from full kick-back on the individual by employing the discovered principles to counter these deleterious effects.

Furthermore, we are living in a society that is "application-mass happy." That is to say, there is a great deal of attention toward procurement of the application mass (such as electronic products, mechanical gadgets, and so on) and its avid consumption, to a point of high reverence and even worship for its existence, with a general neglect for understanding of the postulates or the viewpoint that created them in the first place.

We, as beings, make the habitation of the physical universe a more tenable idea and make it a more favorable universe through the use of physics/engineering principles discussed in this book; for example, by using electricity to tame the savage forces of this universe. Moreover, we use the technical superiority that engineering offers, in order to circumvent or to prevent ourselves from becoming a complete target of the unthinking and reactive forces of this universe.

We use these same very exact rediscovered postulates along with their exact natural laws to counteract the unwanted forces by building precautionary measures against them, and thus, bring about a higher potential of survival for ourselves and our progeny. So our relentless effort to understand the physical universe, even though at first glance appears to be out of curiosity, is actually due to the enormous necessity and urgency of mankind's major tendency toward "survival" as a species.

"Do not lose hold of your dreams or aspirations. For if you do, you may still exist but you have ceased to live."

— Henry David Thoreau (1817-1862)
American Essayist, Poet and Philosopher

The Primary Factor

"Ignorance is the primary source of all misery and vice."

—Victor Cousin (1792 - 1867)
French Educational Reformer, Philosopher and Historian

A certain number of vital facts about the physical universe and its many subsets, such as the universe of waves, electricity, etc., have emerged that make the world of sciences or our universe as a whole, a more understandable and thus, a less puzzling one.

The primary factor intertwined in all aspects of the physical universe is considerations, considerations, and more considerations. The number of considerations built into the physical universe approach infinity in sheer numbers.

By **consideration**, we mean mentally looking at something to create a thought about it; or more specifically, directing the mind toward something in order to make a thought, a postulate, or a decision about its location, function, condition, etc.

These considerations appear invisible at first glance but reign in all aspects of the construction of the physical universe. Every known scientific fact about the physical universe is actually a consideration and cataloguing them in a list format would fill many volumes of books.

However, a few examples of what we mean by considerations could be elucidating. One could think of numerous examples including such things as considerations about the amount of charge or mass of an electron, the size and mass of a proton or neutron, the exact orbit and revolving speed of an electron around the nucleus, the mass and size of nucleus, the amount of mass of an atom, the properties of an atom, the size and shape of a molecule, the density of a material, the size of a planet, the mass and constituents of a star, the size, shape and revolving speed of a galaxy, the distance between galaxies, number of galaxies in the universe, and so on ad infinitum. This would be an endless list.

"Consideration for others is the basic of a good life, a good society."

—Confucius (551-479 BC)
China's Most Famous Teacher, Philosopher, and Political Theorist

The Perceivable Physical Universe

"It is easier to perceive error than to find truth, for the former lies on the surface and is easily seen, while the latter lies in the depth, where few are willing to search for it."

—Johann W. von Goethe (1749-1832)
German Poet and Novelist

Descending from the level of considerations, the whole physical universe (or any part or sub-region thereof) consists of a created space with a certain amount of energy placed in it. This is the perceivable physical universe, which we experience on a daily basis while living on earth.

The energy can be subdivided into *light particles* (such as photons and electrons) and *heavy particles* forming particles of matter (such as atoms and molecules). Particles of matter are obtained by a condensation of lighter energy particles.

Since the temperature is always above absolute zero, therefore energy in the physical universe is in constant motion and has a wave property associated with it.

Furthermore, matter can be considered to be a ridge of energy or a standing wave, which is floating in time. Moving matter also has a wave associated with it according to de Broglie's theory.

Therefore, we can conclude that since all of the energy and matter in the physical universe is in constant motion, we are dealing with waves as the root and source of the entire physical universe. The motion of waves brings forth a time stream.

"Every person is a God in embryo. Its only desire is to be born."

—Deepak Chopra (1947-)
Indian Physician and Author

Moreover, the whole physical universe with all of its gigantic galaxies, billions of stars and billions more planets can be considered as standing waves floating on a relatively timeless basis (i.e., forever), existing in a space, which was created earlier in time than energy or matter.

The time frame of existence of these galaxies and planets is on the order of trillions of years, which vastly exceeds man's length of existence as a species on earth and his continued survival since his appearance less than 100 thousand years ago.

"It is wrong to think that misfortunes come from the east or from the west; they originate within one's own mind. Therefore, it is foolish to guard against misfortunes from the external world and leave the inner mind uncontrolled."

— Buddha, '*The Enlightened One*' (563-483 BC)
The Title of Indian Prince Gautama Siddhartha, Founder of Buddhism

The Inherent Uncertainty

"Maturity of mind is the capacity to endure uncertainty."

— John Finley (1797-1866)
American Poet and Editor

There is an inherent uncertainty associated with any elementary particle of matter. This uncertainty is explicitly expressed by the Heisenberg uncertainty principle. In simple terms, this principle basically states that particles of energy move with such high speed and with such chaos and confusion that it is difficult to determine their position, speed, and energy accurately or with a high degree of certainty.

Because every basic element of matter is made from these uncertain and chaotic particles, every piece of matter perforce remains in an uncertain state. The certainty that we observe on a larger scale (classical physics) is an apparency, which is very transient and rapidly fleeting in time. The certainty that we take pride in measurement at the classical level, is an apparency and not an actuality. The actuality is that we are dealing with particles that exist in such a state of massive chaos and obsessive motion that makes it impossible to determine their nature conclusively.

On a macro-scale, we are dealing with and analyzing a body's behavior and its response to an electrical, magnetic, or mechanical stimulus, by using exact known laws (such as Maxwell's equations or Newton's laws, etc.).

Through the use of elegant mathematical equations, the results of the analysis is usually expressed with a high degree of certainty and accuracy, which makes them look totally plausible and logical; yet, the analysis on this larger scale is merely the study of an average of many uncertain particles that comprise that object. So our analysis of the situation is limited in scope. It is dependent upon and is only meaningful within the framework of the average behavior of many particles and their cohesion to form the object. Furthermore, because these chaotic particles are cohered together, it appears that their uncertainty is masked and is actually reduced to an observed certainty.

At the classical level, there is a particle and there is a wave, which are distinctly separate concepts. A particle (such as an atom or a charged pith ball) consisting of a number of condensed energy particles, is localized and has an exact predictable behavior in space when placed in a force field.

On the other hand, a wave (such as light, sound, etc.) is a form of energy in vibration, which is moving and thus, is not localized. Thus, waves and particles are opposite concepts and in fact dual of each other. The concept of duality between a wave and a particle, is actually the concept of duality between “matter" and "energy.”

At the quantum level, a particle such as an electron behaves like a wave and a wave such as light behaves like a particle (i.e. photon). At this juncture, we are dealing with our shattered certainty of:

a) A particle is no longer a particle but behaves like a wave!
b) A wave is no longer a wave and behaves like a particle!

The whole field of study has suddenly shifted. We have to disbelieve our classical level laws and formulas and study probability of existence of particles and occurrence of events.

Therein lies the answer to this enigma. To reconcile the contradictions regarding our observations of the physical universe, we need to consider the two main facets of existence on a physical plane and realize that we are dealing two simultaneous universes: a) Microscopic Universe, and b) Macroscopic Universe.

While the "Microscopic universe" is the underlying universe for the entire physical universe, the "macroscopic universe" provides the reality level that we experience at every moment of time in our daily existence.

The keynote of the microscopic universe is "Inherent Uncertainties" while that of the macroscopic universe is "Perceivable Illusions," and we are caught in between these two enigmatic universes, like "a rock and a hard place" used as a metaphor, to describe mankind's precarious existence on this planet.

"Quantum mechanics is very impressive. But an inner voice tells me that it is not yet the real thing. The theory yields a lot, but it hardly brings us any closer to the secret of the Old One. In any case I am convinced that He doesn't play dice."

— Albert Einstein (1879-1955)
German Born American Physicist, Nobel Prize in 1921

The Solution to Uncertainty

The uncertainty inherent in the physical universe, which surrounds us at every aspect of our existence on a physical plane, has a dichotomy and that is the certainty in the viewpoint at the thought level.

"Information is the resolution of uncertainty."

—Claude Shannon (1916-2001)
American Mathematical Engineer; Father of Information Theory

This means that at the thought level, the viewpoint can assume a position of total certainty and understanding (the dual concept), which dissolves and replaces the uncertainty and confusion of the physical universe.

In other words, certainty can be postulated into existence so as to make uncertainty junior in importance!

"In the midst of movement and chaos, keep stillness inside of you."

—Deepak Chopra (1947-)
Indian Physician and Author

⊶ ❦ ✶ ❁ ✶ ❦ ⊶

The Governing Postulates

"To rule is easy, to govern difficult."

—Johann W. von Goethe (1749-1832)
German Poet and Novelist

There are three main governing postulates in the physical universe that have a total effect on every aspect of it. These three are referred to as the "*Original Postulates*," which were made very early, at the blue print stage of the physical universe, thought out sequentially but implemented simultaneously, as follows:

A) **THE ORIGINAL POSTULATE #1- EXISTENCE OF SPACE:** This is the first postulate made very early, at the beginning of the physical universe.

B) **THE ORIGINAL POSTULATE #2- EXISTENCE OF ENERGY (or Force)**: This is the second postulate made after the first postulate of the physical universe and gave birth to the first particle of energy and many more soon to follow. In simple terms, it can also be interpreted as: let there be force. This is the concept of universal force that Michael Faraday (an English scientist, 1791-1867) was referring to in his works. This original postulate encompasses the existence of the general concept of force in all of its ramifications and subdivisions, such as waves, fields, radiation, work, power, force, voltage, momentum, charge, current, etc.

C) **The *ORIGINAL POSTULATE #3*- EXISTENCE OF CHANGE:** This third postulate followed right after making the above two original postulates and gave birth to the first particle of matter and the very first few seconds of mechanical time many trillions of years ago.

> In simple terms it says: *Let there be a change or alteration in all aspects of what was created by the earlier postulates.*
>
> This would mean a change in the position of the energy particles or force vectors, a change in the authorship or other aspects of the initial two original postulates, and many other uncountable alterations. This postulate implies the existence of change of location in a created space obsessively on a postulate level, which instantly leads to the concept of mechanical time.

It is interesting to note that the third original postulate, being left on auto-pilot of obsessive change leads to an ever increasing value for the "entropy" of any closed system. The entropy of a system is defined as: *A measure of the degree of disorder, randomness, and uncertainty in a closed system, caused by the random change of particle's location in a created space, or more generally, by the random alteration of data, facts, events, location, authorship, etc.*

Thus we see that the physical universe is set on an automatic increase of entropy, but we can also increase it for any situation or a condition in life unknowingly or causatively due to a lack of confront and an inability to solve a problem "*uniquely.*" For example, by adding random arbitrary data, white lies, fabricated materials, harmful half truths or untruths we would increase the degree of confusion and uncertainty in that life situation or condition, leading to an additional stress and eventual unhappiness for self and others. *Establishing the truth will eliminate the added entropy and reduce it to its original value.*

Of course, this last postulate when put into effect immediately clouded the nature and identity of the postulator and added a whole new dimension to the word obfuscation.

"An idea that is developed and put into action is more important than an idea that exists only as an idea."

— Buddha, '*The Enlightened One*' (563-483 BC)
The Title of Indian Prince Gautama Siddhartha, Founder of Buddhism

Our Relentless Quest for Answers

"At the center of your being you have the answer; you know who you are and you know what you want."

—Lao Tzu (600 -531 BC)
A Chinese Philosopher

At this stage of development, the physical universe is a created universe, which invites us almost on a constant basis to explore in order to discover its very many intricacies or facets of existence: on a microscopic scale (e.g., new subatomic particles, atoms, molecules, etc.) or on a macroscopic level (new phenomenon, planets, stars, galaxies, etc.).

We human beings are on this relentless quest to analyze and unravel the many "secrets" of this created universe without ever stopping for a moment to visualize the bigger scheme of affairs, which includes the realization of the fact that before there was an electron or an atom there was the thought of an electron or an atom. Before there were these massive galaxies, stars and planets, there were the thoughts of galaxies, stars and planets. Thus, the actual things to explore are the considerations or the thoughts behind our created universe, and not the created objects themselves as is done endlessly in sciences.

On a smaller scale, it is not difficult to see that before there was the New York city or the Empire State building, there was the thought of New York city and the Empire State building. We get so fascinated by the resultant application mass of these powerful thoughts (expressed in terms of precise postulates) that we forget about the simple mechanics of creation of a universe. The sheer number of postulates, sub-postulates and considerations built into the current physical universe, is appalling. It is an enormously large number, which actually approaches infinity. The number is actually so large that it is unfathomable!

Of course, it goes without saying that once we have discovered the actual postulates concerning any portion of the physical universe, it is only a matter of time before one derives or discovers a series of governing and irrefutable laws that will completely describe the subject matter fully, and will delineate the behavior of all of the related particles, elements, objects, etc., that would supplement the postulate portion.

In the discovery process, one searches for the thought or postulate behind a certain phenomenon under consideration. Once the exact thought behind a phenomenon is isolated and fully known, the postulates would soon emerge and thus, be *recovered.* Only then can they be stated concisely.

The next logical step in this sequence would be the derivation of all of the axioms, natural laws, rules and theorems by mathematical proof or by accurate observations. This process has shown to be true over and over in the scientific arena. It is a fascinating principle to remember when dealing with sciences.

"Judge a person by their questions, rather than their answers."

— Voltaire (1694-1778)
French Philosopher

Our Complex World

"To be yourself in a world that is constantly trying to make you something else is the greatest accomplishment."

—Ralph Waldo Emerson (1803-1882)
American Essayist

The complex technological world surrounding us in every aspect of our existence has come about and is founded upon some basic simplicities.

Starting with the vital element and the basic considerations, we obtain the microscopic and macroscopic universes stacked up one on top of the other, followed by the scientific arena of modern sciences, which have created enormous technical application mass through the years.

The Figure on the next page shows how the viewpoint postulated our complex world into existence, starting with the microscopic universe to the macroscopic universe leading to our modern sciences and their empowering technical application mass (almost an infinity of visible application mass), created to benefit all mankind.

"We are what we think. All that we are arises with our thoughts. With our thoughts, we make the world."

— Buddha, '*The Enlightened One*' (563-483 BC)
The Title of Indian Prince Gautama Siddhartha, Founder of Buddhism

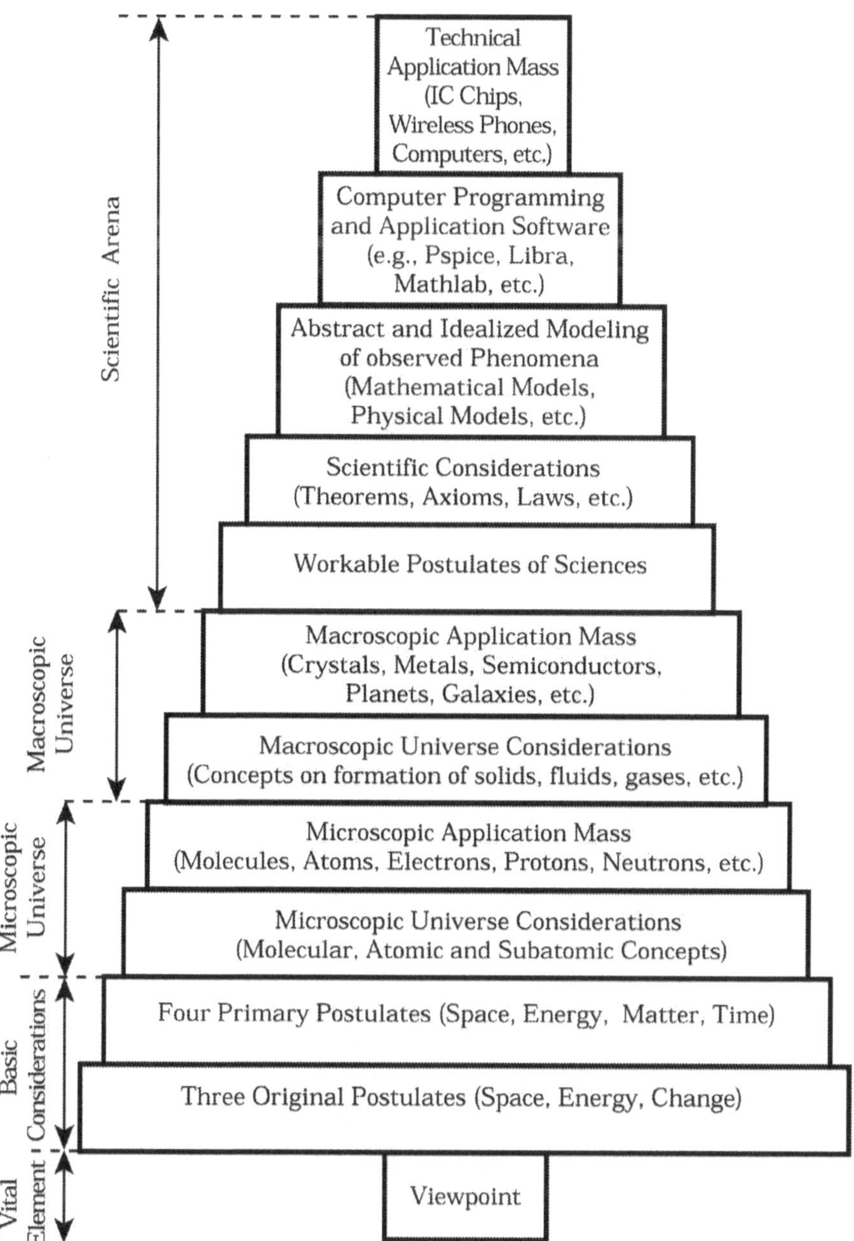

From The Viewpoint To Our Complex World

The Significance of Electricity

"I shall make electricity so cheap that only the rich can afford to burn candles."

—Thomas Alva Edison (1847-1955)
American Inventor, who held a world record 1,093 patents,
creator of world's first industrial research laboratory

The discovery of invisible electronic waves has been one of the greatest scientific triumphs of the nineteenth century science in the field of electricity and with it a whole new vista opened up—a whole new vista of monumental discoveries of unimaginable proportions, which has put us on a whole new course of activity and plateau of existence.

Having risen intellectually and technically above other creatures of this planet, we have an inherent right to know the answers to the puzzles and mysteries that we encounter in our environment. This right is no more than having the "right key" to the locked doors of knowledge. The metaphorically speaking "right key" will open the locked doors of knowledge that have been locked from inside for centuries, if not thousands of years. The ancients tried to open these locked doors, but their success was limited because they lacked an essential factor—electricity and its companion magnetism.

Electricity as a subject has been wrapped up in an aura of magic and mystery through the years, and even today most people regard it a very difficult subject, filled with technical terms and mathematical expressions, which derails most of the non-technically oriented public and thus, portrays an untenable and incomprehensible subject.

Truth be told, electricity is an extremely vital subject to life and existence since life organisms make great use of it in not only carrying out daily physical tasks using nerve cells to send it to different organs (Bioelectricity) but also it is employed in imagination as well as observation of the physical universe and its subsequent recordation, storing and playback of the recorded information on a mental level.

There is no monopoly on knowledge and it belongs to those who reach for it; therefore, we have the right to know electricity well, well enough to use it ethically and apply it competently to the world around us.

We need to know the basics of this vital science in simple and understandable terms so that not only the puzzlement of life ceases and becomes solved, but also through its intelligent application to ourselves and our fellow man, we will produce stellar results in many related fields (Obviously, such intelligent and ethical applications will exclude crude and brutal utilization of raw electricity to Man, as in electric shock therapy, etc., currently used as a "healing method" and passed off as "treatment!")

Therefore, it could be said that having discovered electricity in its many varied forms, we, as a species, are being rewarded amply on a daily basis for our awareness and understanding of this invisible and finer form of energy (called electricity) through the many inventions and the attendant technological innovations.

"Electricity is like faith. You can't see it, but you can see the light."

—Anonymous

The Evolution of Sciences

Any physical science evolved from basic simplicity and gave birth to very complex technical application masses.

With the help of a postulating viewpoint (or simply a viewpoint) and the three powerful original postulates in place, which leads to four primary postulates, we come into possession of a microscopic universe. This in turn leads to the creation of the macroscopic universe, which forms the backbone of our life's reality and existence.

These two interwoven universes along with their corresponding application masses provide the life blood of the scientific considerations, laws, hypothesis, theories and technical investigations leading to the life-enhancing application masses, which we triumphantly announce as the latest technology!

Understanding the main concepts as well as their limitations and frailties of sciences makes us become aware of their dependency on the early postulates of our universe, and we come into possession of a tremendous mount of simplicity about the way their theories and philosophy of operation have stacked up through the years.

Keeping this point in mind, we are well on our way to understanding the "ultimate confusion" we commonly refer to as "the physical universe" as shown on the next page.

"The true sign of intelligence is not knowledge but imagination."

—Albert Einstein (1879-1955)
German Born American Physicist, Nobel Prize in 1921

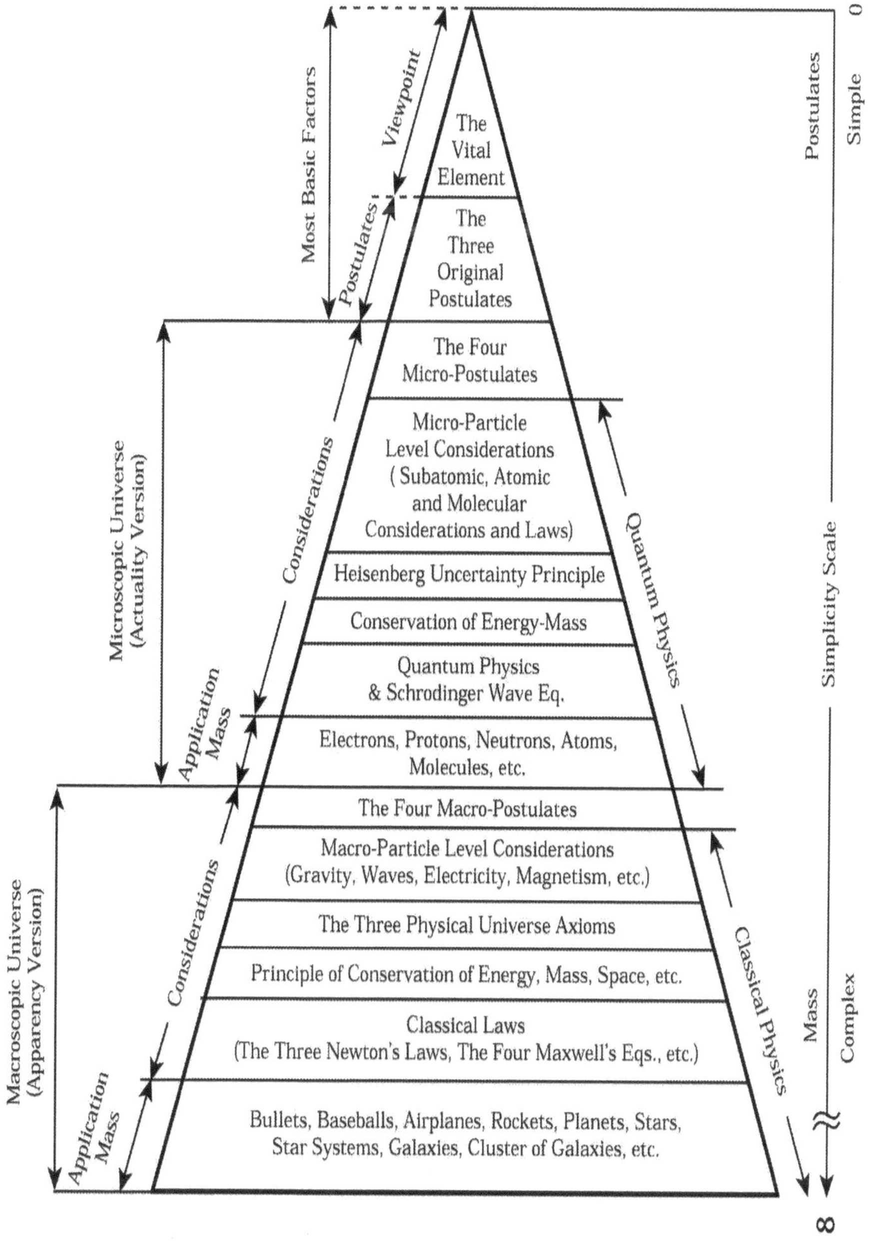

The Vital Element

The Three Original Postulates

The Four Micro-Postulates

Micro-Particle Level Considerations (Subatomic, Atomic and Molecular Considerations and Laws)

Heisenberg Uncertainty Principle

Conservation of Energy-Mass

Quantum Physics & Schrodinger Wave Eq.

Electrons, Protons, Neutrons, Atoms, Molecules, etc.

The Four Macro-Postulates

Macro-Particle Level Considerations (Gravity, Waves, Electricity, Magnetism, etc.)

The Three Physical Universe Axioms

Principle of Conservation of Energy, Mass, Space, etc.

Classical Laws (The Three Newton's Laws, The Four Maxwell's Eqs., etc.)

Bullets, Baseballs, Airplanes, Rockets, Planets, Stars, Star Systems, Galaxies, Cluster of Galaxies, etc.

Most Basic Factors

Viewpoint

Postulates

Considerations

Application Mass

Considerations

Application Mass

Microscopic Universe (Actuality Version)

Macroscopic Universe (Apparency Version)

Quantum Physics

Classical Physics

0

Postulates

Simple

Simplicity Scale

Mass

Complex

∞

The Evolution Of Science And Their Complex Byproducts From Basic Simplicity.

PART XIII

The Unraveling of the Code

The Future Universe

"Let your hopes, not your hurts, shape your future."

—Robert H. Schuller (1926-)
American Minister, Entrepreneur and Author

"The Future Universe," which is constantly being shaped or carved out of the current physical universe by the beings who occupy it, will be a far more "dynamic and gentler" universe. We have and are surely developing a new universe with the force of our reasoning, a force which this universe has no concourse with and has never seen before.

With every revolution of earth we take one more step in the direction of creating this universe. It will be much superior to the present one, which we have been observing for the last 100 thousand years as a species of human beings.

Needless to say, the above statements exclude all of the societal psychosis and irrationalities of Man, such as wars, genocides, terrorism and other such ills, which can be classified under the heading of "Man's inhumanity to Man."

From this brief examination of "the physical universe" and "the universe ahead" we can see that the application mass (whether, technical, generalized or personalized) of a universe is only the visible embodiment of the postulates of the viewpoint(s). There are a myriad of considerations that affect the size, mass, density, composition, location, geometry and other aspects of the created things. The collective mass of all created forms and things is what we refer to as the application mass of the universe. They are the final receipt point or the effect of all of the postulates put into practice and as such they are of junior importance!

Therefore, we can observe that any portion of the mechanics or any component thereof bears a mental stamp of the postulate that created it and placed it there. Oftentimes when one looks at a universe and sees the panoramic view of the created things, one may wonder which is more important: the "created thing" or the "postulate" behind it. Knowing this material will help greatly in solving this dilemma expediently.

In conclusion, the physical universe, as a total-effect type of universe, being a total byproduct of the original postulates, has a future which is proportional to our complete understanding of its postulates and its many derived features.

The upcoming physical universe is a direct function of how well we understand the postulates clearly laid out in this book as well as our own relationship with regard to it and the role that our thoughts play in its existence.

Furthermore, how successfully we tame its many savage forces and put them to use toward enhancing our own survival, while harnessing our own destructive impulses in order to prevent its ultimate demise, will have a huge bearing on the fate of the earth, the planet we call our home at the moment in this forlorn corner of the universe*!*

"Two things inspire me to awe -- the starry heavens above and the moral universe within."

—Albert Einstein (1879-1955)
German Born American Physicist, Nobel Prize in 1921

Beyond the Physical Universe

As one looks across the aggregate of modern physical sciences, one cannot help but notice that the bulk of these sciences are held up by considerations. These considerations are supported in part by a series of powerful postulates made by the founders and pioneers of the subject, which have shaped the entire subject as a living and growing entity. Of course, the postulates are furnished by a viewpoint or a composite viewpoint, and are developed through the many centuries that physical sciences have formally existed.

These postulates are capable of changing the existing state of a science and/or creating new areas of study. In actual fact, the postulates trickle down through natural laws, theorems, design techniques, charts, and tables to the final byproduct which is the application mass.

Conversely, one can follow this line of logic and observe that before an application mass can exist, there must be a manufacturer, before that a designer, and before that a postulator of that application mass; and before all of these, there *comes a viewpoint that exists exterior to this process and is solely concerned with observation and postulation of new and orderly thoughts*, which reflects nature and its inner workings.

Thus, beyond the physical entity of a product or an object lies a higher entity called a viewpoint that through a mental process arrives at the postulates, and eventually makes it possible to create the application mass. The key postulates to be made in any science start with some basic postulates concerning space, energy, matter and time. These are first stated purely as concepts, which then materialize as forms (see the Figure on the next page). These postulates, even though initially made on a sequential basis, are implemented simultaneously in actual practice.

Moreover, any extant application mass carries with it a physical stamp of the simultaneous implementation of the basic postulates and has all of them built into its constituent components at each moment in time. The dual of the postulating viewpoint is the observing viewpoint as shown below.

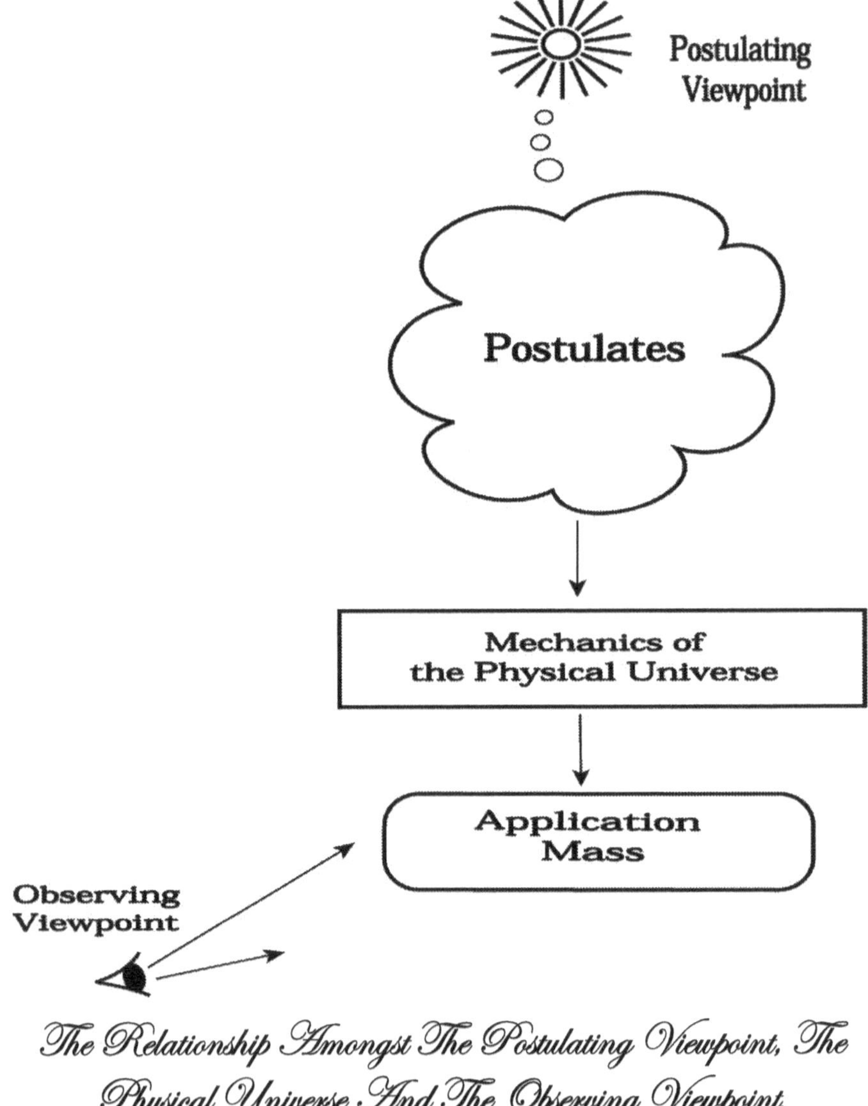

The Relationship Amongst The Postulating Viewpoint, The Physical Universe And The Observing Viewpoint.

"Just as our eyes need light in order to see, our minds need ideas in order to conceive."

—Napoleon Hill (1883-1970)
American Author

So one can see that beyond the physical universe lies thought in its purest form: an "*analytical and aware thought,*" that has the ability to create postulates, form concepts, create space, create energy, and matter, and put them in motion as an application mass and thus, create a time stream as shown below.

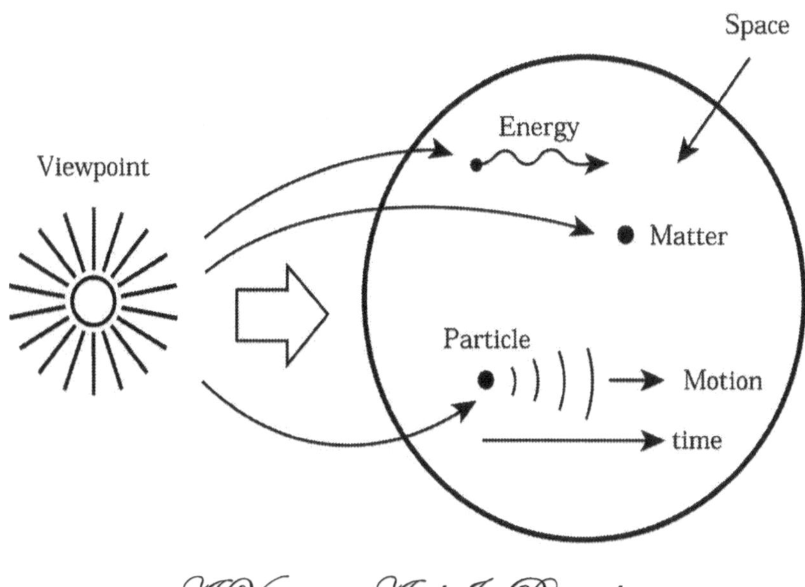

A Viewpoint And Its Byproducts.

"Unless you try to do something beyond what you have already mastered, you will never grow."

—Ralph Waldo Emerson (1803-1882)
American Essayist

It is important to note that we are assuming that the postulated subject is a workable science. Failing that assumption, the viewpoint can be a confused or uninformed entity generating a totally unworkable and unproductive subject leading to no or false application mass. Examples of such unworkable subjects include superstition, voodoo, sorcery, witchcraft, fascism, communism, psychiatry, psychoanalysis, sociology, psychology, shamanism, politics, and so on ad infinitum.

The aforementioned are very subjective branches of study filled with arbitrary factors of all kinds. Furthermore, this list of unworkable subjects should come as no surprise to the reader, since the truly workable sciences are few but all are axiomatic in nature, have a well-defined nomenclature, and are founded upon a set of very precise postulates.

"The world is so empty if one thinks only of mountains, rivers and cities; but to know someone here and there who thinks and feels with us, and though distant, is close to us in spirit - this makes the earth for us an inhabited garden."

—Johann W. von Goethe (1749-1832)
German Poet and Novelist

Analyzing The Code

By Code we mean *any unique set of data or combination of symbols, signals, actions, etc., that will open a passageway to a hidden location (such as a vault), or unlock the pattern of behavior of some puzzling and mysterious data, thus, revealing the information behind the mystery.*

Cracking the ultimate physical universe code is **a process of discovery** and one, perforce, must travel the opposite direction to the laid out pattern of the pyramid of knowledge. It actually requires the complete understanding of the inner workings of the underlying parts of the code.

To break apart the main code requires us to crack six underlying and yet separate codes, which are the essential ingredients and are vital if one truly desires to achieve the answer. These six separate codes when summed up give us the final code. They can be briefly stated as:

I. *The Application Mass Code (also called the Mass Code),*

II. *The Model Code (also called the Imitation Code),*

III. *The Formulation Code (also called the Hypothesis Code),*

IV. *The Wave Code (also called the Energy Code),*

V. *The Pattern Code (also called the Arrangement Code), and*

VI. *The Postulate Code (also called the Thought Code).*

"We're given a code to live our lives by. We don't always follow it, but it's still there."

—Leonard Gary Oldman (1958-)
British Actor

The Figure below shows the sequence of codes needed to be cracked before the ultimate code to the physical universe is cracked.

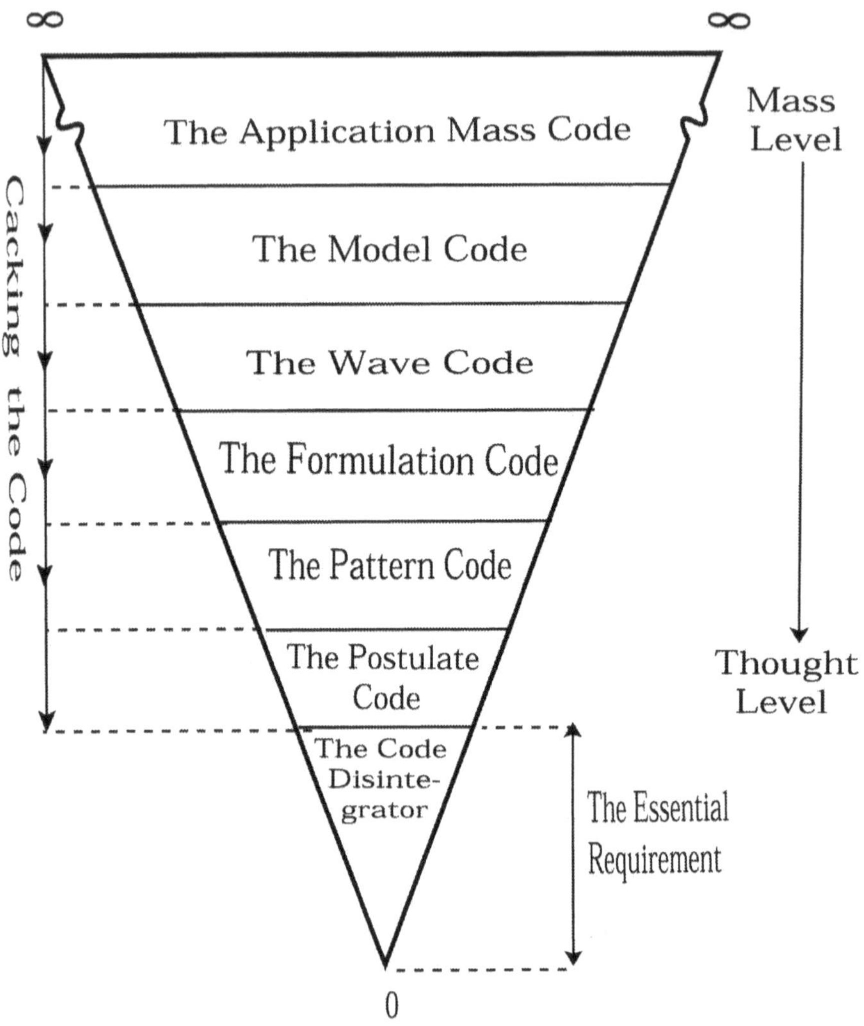

To Crack The Ultimate Code Requires Cracking Six Separate Codes Sequentially.

Cracking The Application Mass Code

"The true mystery of the world is the visible, not the invisible."

—Oscar Wilde (1854-1900)
Irish Poet, Novelist, and Dramatist

By application mass, we mean an object, a phenomenon or any perceivable thing that one is trying to identify and crack its code in the process and to thereby gain total causation over it.

To crack the code of any intended application mass requires us to start with the observation of the application mass and the visible or perceivable parts of its apparent manifestations. This initial step is purely an observation process, an extremely important and powerful step, which, if its code is properly deciphered and cracked, lays the groundwork for the next several codes to be cracked!

In the first stage of this process, we need to observe and examine the application mass. In the next phase of this process, we look for all of the characteristics and attributes of the application mass. Next we prepare a complete set of documents describing each aspect of the thing or phenomenon, which clearly explains and reveals all of the hidden aspects and manifestations in very simple language.

The code to the application mass is cracked when we have established exactly and specifically what it is that we are studying, in terms its space, its energy level, its accompanying mass, and the length of time of its existence. It includes a complete observation and examination of the shape, size, density, color, material texture, etc., of the intended application mass.

For example, circa 600 BC the Greek philosopher, Thales, made an initial observation about the phenomenon of static electricity but did not crack its code at that time. Mankind had to wait 2200 years for William Gilbert, a royal physician in England, to study the phenomena, make meticulous observations about this incredible energy form, and finally cracking its code in 1600 AD. He went on to gain worldwide fame for this discovery and made it possible for others to continue this process and thus, crack other subsequent codes. He was the first to call it "*electricity*," which to date still bears this name.

"Life would be much easier if I had the source code."

—Anonymous

Cracking The Model Code

The model code requires us to simulate an observed application mass, whether it is an object (e.g. a TV) or a phenomenon (such as wave propagation).

The simulation process consists of making a model, which could be in the form of a) a mechanical object (made out of wood, clay, etc.), b) an electrical circuit, c) a mathematical formula, d) a step by step procedure, or e) a virtual object generated by a computer, which simulates and represents the application mass on an electronic level.

Even though rare, a mental representation existing purely on a thought level could serve just as well as a model code. However, there is an inherent problem with this mental model, since it can not be shared with or shown to others for the purposes of communication or proper edification. Thus it is an individual research and not a generally accepted method of simulation of a phenomenon.

For example, Nikola Tesla (a U.S. inventor, 1856-1943) had made a mental model of alternating current and the AC machine, long before cracking its final code on a physical plane. Eventually, he had to consolidate his mental models physically before his concept of alternating current and his brilliant inventions could gain wide acceptance and be agreed upon generally.

"By three methods we may learn wisdom: first, by reflection, which is noblest; second, by imitation, which is easiest; and third by experience, which is the bitterest."

—Confucius (551-479 BC)
China's Most Famous Teacher, Philosopher, and Political Theorist

This step is an important step in further understanding of the application mass. Building this model, if done correctly, will go a long way in making the application mass and its behavior a solid reality in one's mind as well as in the physical universe!

Another example is James Clerk Maxwell (Scottish physicist, 1831-1879), who built a mechanical model of the electric and magnetic fields out of small spheres placed in a hexagonal structure, and showed how the electromagnetic waves propagate. Through the investigation of this physical structure, he was finally able to crack the model code of electronic waves successfully and actually formulated the wave behavior mathematically, more of which in the next code.

In our modern times, most of the model codes are done electronically, that is, via computers and the related software that describe the behavior of the intended application mass. These models so generated could be called virtual models, since they do not involve building of physical objects, and yet they provide an efficient method to build a replica in a short period of time with enormous flexibility and ease and thus, crack the model code a lot faster!

"The purpose of science is not to analyze or describe but to make useful models of the world. A model is useful if it allows us to get use out of it."

—Edward de Bono (1933-)
Maltese Writer, leading authority in field of creative thinking

Cracking
The Formulation Code

To crack the formulation code, one needs to be well versed and utterly familiar with the prior two codes and have a full understanding of both, before proceeding to crack this next code.

This is a highly abstract level code, which requires tremendous amount of mental agility and prowess. One needs to piece together "the application mass code" and "the model code" and come up with a formula (mathematical or otherwise) or a hypothesis, which describes the application mass with enormous precision and can predict the phenomena at least on paper. When this is done, the code is considered to be cracked.

For example, James Clerk Maxwell having developed the model code for the electromagnetic wave propagation, went on to formulate his famous "Wave Theory," which universally describes all electrical phenomena and predicts the existence of any and all electromagnetic waves in space at any point in time, far in advance of its actual discovery—clearly an enormous achievement!

This mathematical feat, which at the time was purely an unproven hypothesis, had yet to wait another two decades to pass the wave test by another scientist. He gained worldwide fame for the discovery of this code but his short life did not allow him to go on to crack "the wave code," or "the postulate code," so those tasks ware left to others such as Hertz, Einstein, etc. years later!

"The formulation of a problem is often more essential than its solution, which may be merely a matter of mathematical or experimental skill."

—Albert Einstein (1879-1955)
German Born American Physicist, Nobel Prize in 1921

Cracking
The Wave Code

To crack the wave code requires subjecting the formulation code to an exhaustive series of experiments, which beyond the shadow of doubt proves or disproves the formulation code discovered by the predecessors.

In this step, all we are trying to do is to throw a wave at the application mass and examine the outcome. We start with applying a wave (whether electrical, mechanical, etc.) into or onto the application mass and record the results thus obtained. Then we may change the frequency of the wave or its amplitude or its angle of incidence and perform many such tests and thus, gather an enormous amount of data by doing this step diligently.

"When I have fully decided that a result is worth getting, I go ahead of it and make trial after trial until it comes."

—Thomas Alva Edison (1847-1955)
American Inventor, who held a world record 1,093 patents, creator of world's first industrial research laboratory

The wave code is considered to be cracked when the obtained results are accurately tabulated and compiled into understandable tables and charts, all the while preventing the unadulterated data to enter the measurement process by making sure ***no wild variables*** *have entered into the experimental process. The pure results so obtained further shed light on the behavior of the application mass.*

Cracking this code requires a total awareness of the experimental environment, the presence input signals, parasitic inputs, output signals, spurious outputs and the existence of wild or uncontrolled factors and parameters, etc. It requires devotion to the truth and complete determination to find it regardless of personal bias or vested interest!

Furthermore, as part of cracking of this code, there should be no such thing as an exception to the formulation code or else it isn't!

Having cracked the wave code, we may regard our understanding to be on safe grounds as far as the application mass is concerned!

An example of cracking the wave code is "Maxwell's Wave Theory" by Heinrich Hertz, twenty years after Maxwell proposed its formulation code. Hertz meticulously performed many wave experiments and compiled the results into many graphs and tables using accurate test conditions that prevailed at the time, leaving all wild variables out of the test results.

Through the unadulterated results that he had obtained, he cracked the wave code by showing how the formulation code applied, thus dispelled the doubts about the validity of the formulation code. He brought a whole new level of observation and understanding to the wave phenomena, based upon factual data that no one had been able to produce for years.

"The secret of success is to be in harmony with existence, to be always calm, to let each wave of life wash us a little farther up the shore."

—Cyril Connolly (1903-1974)
English Critic and Editor

Cracking
The Pattern Code

"A cloud does not know why it moves in just such a direction and at such a speed...It feels an impulsion...this is the place to go now. But the sky knows the reasons and the patterns behind all clouds, and you will know, too, when you lift yourself high enough to see beyond horizons."

—Richard Bach (1936-)
American Writer, Author of "Jonathan Livingston Seagull"

To crack this code one needs to compare the data collected from the wave code to the answers obtained through the formulation code. The percentage error between the two should be very small.

One needs then to classify all of these patterns into a series of recognizable patterns and obtain rules of thumb to speed up the prediction process visually as well as to provide a quick grasp of the different aspects of the application mass.

By using exact patterns, which are well documented in charts, graphs, diagrams, and other visual indicators, one can gain quite a bit of understanding about the behavior of any new phenomenon that presents itself for study, without having to rigorously delve into or use the formulation code.

In fact, often times the formulation code is worked into an easy to use program whereby the operator can predict the behavior of the application mass under many hypothetical conditions without having to resort to tedious and time-consuming calculations.

For example, Heinrich Hertz, having cracked the wave code, was able to proceed to the next level and actually crack the pattern code by comparing answers from the formulation code to that of wave code and find amazing closeness between the two and actually confirm the mathematical prediction of Maxwell that, “Light is also an electromagnetic wave” and thus, a subset of the Maxwell's "Wave Theory." This was a conclusion that was not easily palatable to many optical scientists, but the facts as revealed by the pattern code could not be refuted.

Hertz and his successors were able to prepare many charts and graphs to speed up understanding of the wave phenomena, notable amongst them is the Smith Chart (developed by P. Smith of Bell telephone Labs in 1939) which has revolutionized the way we analyze wave propagation on a transmission line and design matching circuits for maximum power transfer from an electronic wave source to a load.

"Art is the imposing of a pattern on experience, and our aesthetic enjoyment is recognition of the pattern."

—Alfred North Whitehead (1861-1947)
British Mathematician and Philosopher

Cracking The Postulate Code

"All that we are is the result of what we have thought. The mind is everything. What we think we become."

— Buddha, '*The Enlightened One*' (563-483 BC)
The Title of Indian Prince Gautama Siddhartha, Founder of Buddhism

Having cracked all of the previous codes: application mass code, model code, formulation code, the wave code, and the pattern code, we have achieved a very high plateau of mental understanding about the application mass. This understanding, if applied accurately and in combination, should be able to crack the final code: the postulate code.

Cracking this code requires a thorough understanding of what has gone before by the pioneers of the field, a good grasp of the pyramid of knowledge, and how a dynamic thought such as a postulate can create an application mass.

Thus, one is faced at this stage with the discovery of the exact postulate (or sequence of postulates) that has gone into the creation of a certain application mass. Once the exact postulate has been discovered with a high degree of accuracy, the postulate code is considered to be cracked.

"A man's private thought can never be a lie; what he thinks, is to him the truth, always."

— Mark Twain (1835-1910)
American Humorist and Writer

For example, in the author's previous book, *The Gateway to Understanding: Electrons, to Waves and Beyond*, the postulates regarding the "field of electricity" were revealed to consist of six essential postulates from which any and all electrical, magnetic, or electromagnetic phenomena can be derived and all of the pertinent laws could be extracted. These postulates are: a) Charge postulate, b) Electric field postulate, c) Electromotive force (emf) postulate, d) Magnetic field postulate, e) Force postulate, and f) Wave postulate (Please see Ref. #12, Chapter 12 for further details).

Another example, is any piece of matter, which exists in the physical universe. It was shown in the author's previous book that there are only three irreducible postulates that create any piece of matter. These postulates in the order of importance are: a) Space, b) Energy, and c) Eternal change which leads to creation of matter and time (Please see Ref. #12, Chapter 2 for further details).

Of the six codes, the postulate code is the most important code, since it encompasses all of the prior five codes. Once it has been cracked one can then create the application mass at will and thus achieve a level of total causation over that particular application mass. In other words, the application mass is no longer a source of mystery or awe to him and actually has lost its power over the individual!

"The cradle rocks above an abyss, and common sense tells us that our existence is but a brief crack of light between two eternities of darkness."

—Vladimir Nabokov (1899-1977)
Russian Born American Novelist, Critic and Author

Creating the Code to Any Universe

Having understood the six codes that go into cracking the code of the physical universe, we can now utilize the discovered principles to our advantage.

By a simple observation, we can see that we can create the code to any universe by a reversal of the process, as follows:

a) **The Postulate code**: We implement the postulate code, by starting with a thought of what the code should be.

b) **The Pattern Code**: This code is created by developing a series of patterns that further refines the code and needs to be put in place.

c) **The Formulation Code**: To implement this code we develop a series of formulas, procedures, etc., which are precisely laid out with no exceptions and relay exactly what the code consists of.

d) **The Wave Code**: A series of tests and experiments are conducted to make sure the implemented code works and does not malfunction.

e) **The Model Code**: A small replica of the application mass with the code in place is made to bring the concept into its level of reality before the application mass is mass-produced on an assembly line.

f) **The Application Mass Code**: The actual object with the code in place is produced on a large scale. The implemented code is stamped on any and all application mass so produced, whose sheer number could approach infinity.

"Imagination is the beginning of creation. You imagine what you desire, you will what you imagine and at last you create what you will."

—George Bernard Shaw (1856-1950)
Irish Writer, 1925 Nobel Prize for Literature

The Figure below shows how the six levels of code need to be created and implemented before the code becomes a functional code for a given application mass.

It should not be forgotten that the most essential element of the creation of the code is the code creator or the postulating viewpoint, which is essential in this process.

"What we are today comes from our thoughts of yesterday, and our present thoughts build our life of tomorrow: Our life is the creation of our mind."

— Buddha, '*The Enlightened One*' (563-483 BC)
The Title of Indian Prince Gautama Siddhartha, Founder of Buddhism

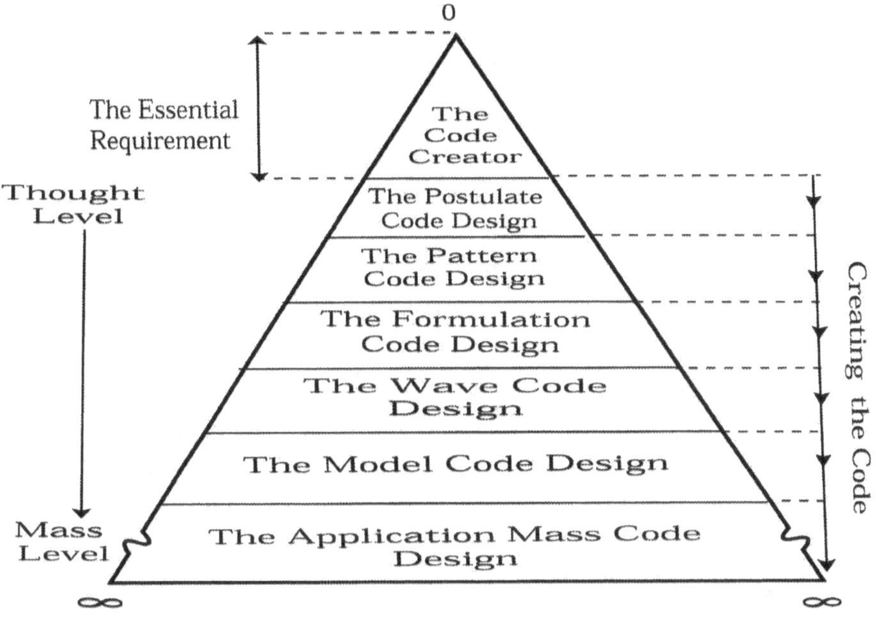

The Process Of Creation Of Any Code In Any Universe.

The Physical Universe Cracked!

The fact that the physical universe never reveals its nature or origin, constantly posing questions for its resolution and daring us constantly to find its ultimate truth, indicates that it can not communicate to us on a mental level. In other words it lacks an ability to communicate because it is an unthinking and unknowing universe.

The physical universe has no knowledge of itself and does not know its whereabouts or to where it is going, since its fate is dictated by the postulates that made it. It has no power to perceive, know, or feel because it is the result of the implementation of certain postulates done by a viewpoint (or collective viewpoints). As such, it only exists on a "total effect" basis.

Because life imbues it with so much of its own energy, it appears to be a live universe at times, but actually it is a dead universe. It appears to be a very intriguing universe by drawing one's attention to it to unlock its many mysteries, yet it offers no help to resolve the posed dilemmas. One cannot establish a two-way communication with it to extract some vital information about its true nature or its formation history.

However, in the absence of a two-way dialogue, it absorbs one's attention and demands obedience. It actually drowns one's beingness and attempts to consolidate one's thoughts into solidity and to pattern them after its many "action-reaction" mechanisms, which are built into it on an automatic basis. It could be classified, by some, as an "attention-trap" type of universe.

In a normal state, it does not communicate but when it does, it communicates on a very gross, physical plane—only through a crude level of force fields. On the other hand, it expects us to use reason and intelligent force when communicating with it or handling any part of its components. A double standard of some sort!

It asks one on a constant basis to solve its many riddles. It only comes to life when life has invested and embedded so much of its own characteristics in it. Only then, does it appear to be alive, artificially and as an apparency! An example of this would be a robot operating on artificial intelligence principles, which appears to be life-like and sentient but it is far from it.

Therefore, due to the communication inability of the physical universe, the knowledge that we extract from it can only come about *on a trial and error basis along the guidelines of the "Scientific Methodology" as well as the utilization of the "Six Basic Codes.*"

A wide observation of the physical universe indicates to us that it exists completely on an "action-reaction" basis and has been in the making for trillions of years. However, the use of the "dichotomy principle" makes our physical universe a more understandable universe because it points out the two fundamental sides to existence: one visible (junior) and the other invisible (senior).

The knowledge obtained about the physical universe through the use of the dichotomy principle would be far superior to one that could ever be obtained from a thorough study of physics or pure observation of the physical universe itself as has been done by scholars in physical sciences for centuries, if not millennia.

Thus, we can see that the principle of dichotomy points out the other side of the coin, which has been neglected for thousands of years (if not more) by Man. It points out that our "beloved" physical universe exists on a "created space, created energy, created mass, and created time" basis and floats around on a totally automatic level of operation.

The resolution of the puzzles of the physical universe is actually half of the answer. This is because the answers obtained from the "created" side, are only half of a larger concept under the heading of "*creating and created*". The exploration of the other half (i.e., creating) is a study in postulates, which has a small subset called "design," commonly used in engineering.

In Conclusion, we can see that *the physical universe, all by itself, does not contain all of the answers to its own resolution, since it is an "incomplete universe." The answers to its creation and current existence lie in the postulate and consideration band, which is a subset of the "thought universe." Moreover, applying the dichotomy principle, with the help of the six codes in actual practice, would yield the desired answers about the universe, and for the first time, a puzzle called "the physical universe" can be considered cracked wide open for anyone to see!*

The Figure below shows our relationship, as viewpoints, with respect to the physical universe's past (the "then physical universe"—*an illusion reality)*, present (the "now physical universe"—*A shared reality)*, and future (the "will-be physical universe"—*an anticipated reality)*

"***Imagination is everything. It is the preview of life's coming attractions.***"

—Albert Einstein (1879-1955)
German Born American Physicist, Nobel Prize for Physics in 1921

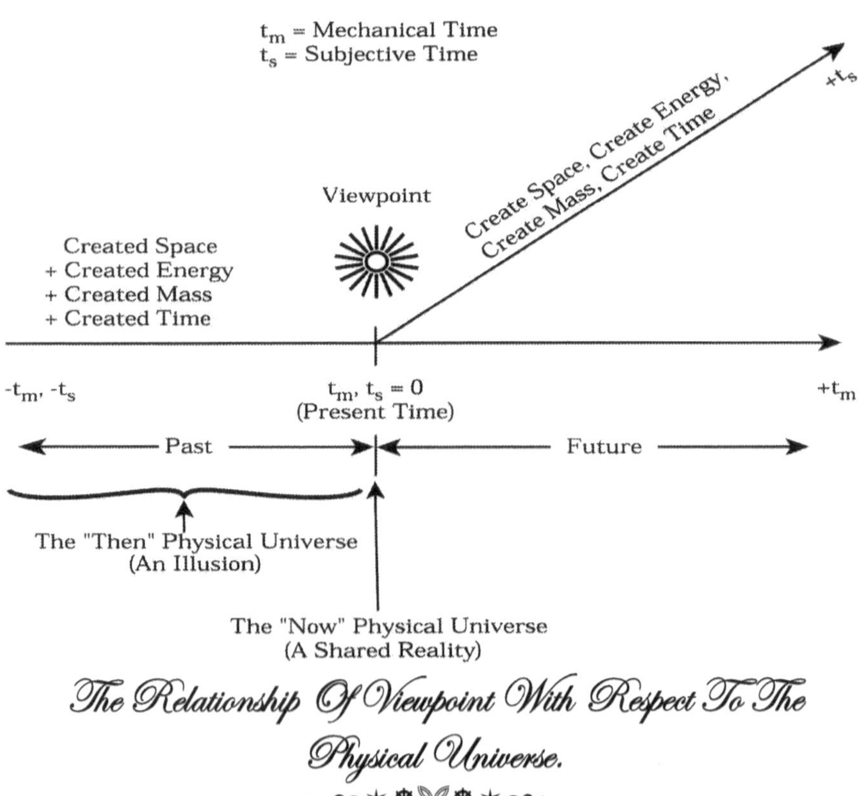

The Relationship Of Viewpoint With Respect To The Physical Universe.

Epilogue

Cracking the code of the physical universe has a special meaning in this work. It does not mean that by finding the code one is then able to obliterate or destroy the physical universe. Quite the contrary, acquiring the combination to the code enables one to soar to new heights of existence and enlightenment, and become cause over the material universe to a point of commanding it and actually observing before one's eyes that his commands are carried out to the fullest extent.

Such a feat and accomplishment requires the presence of several key factors, which have been explained in depth and fully illustrated throughout this work.

In summary, to crack the code of the physical we need the most essential element and that is the code disintegrator. Once the code disintegrator is at hand, then and only then, can we unlock the gate by cracking six separate codes, called the "*Application Mass Code,*" "*The Model Code,*" "*The Formulation Code,*" "*The Wave Code,*" *"The Pattern Code," and "The Postulate Code,"* in that order.

"No one saves us but ourselves. No one can and no one may. We ourselves must walk the path."

— Buddha, '*The Enlightened One*' (563-483 BC)
The Title of Indian Prince Gautama Siddhartha, Founder of Buddhism

Utilizing the code disintegrator one starts by examining the application mass and step by step dissolves it (first code) and winds up at the end with the postulate (sixth code) that made the application mass and thus, cracking the overall code in the process!

This feat is achieved by traveling the pyramid of knowledge in the reverse direction, a journey of discovery and enlightenment, which hands one not only the key to the physical universe but also the blueprint of the key to one's own universe.

Furthermore, by reversing the process, one can create codes for any universe, which is far more powerful than cracking codes!

This book lays out the milestones one needs to achieve in order to arrive at the final code combination, and for you fellow traveler, the journey has just begun since it surely is worth the travel!

"The evolution of knowledge is toward simplicity, not complexity."

—L. Ron Hubbard (1911-1986)
American Philosopher, Writer, Educator, and Humanitarian

Now, You Have The Ultimate Key to Our Universe!

The Real Journey Has Just Begun,

Bon Voyage!

⊶ ♦ ✶ ❁ ✶ ♦ ⊶

APPENDIXES

A. International System of Units (SI)

B. Physical Constants

C. The System of Units

APPENDIX A

International System of Units (SI)

QUANTITY	UNIT	SYMBOL	DIMENSION	TYPE
CAPACITANCE	FARAD	F	C/V	Derived
CHARGE	COULOMB	C	A-s	Derived
CONDUCTANCE	SIEMENS	S	A/V	Derived
CURRENT	AMPERE	A	A	Basic
ENERGY	JOULE	J	N-m	Derived
FORCE	NEWTON	N	Kg-m/s^2	Derived
FREQUENCY	HERTZ	Hz	1/s	Derived
INDUCTANCE	HENRY	H	Wb/A	Derived
LENGTH	METER	m	m	Basic
MAGNETIC FLUX	WEBER	Wb	V-s	Derived
MAGNETIC INDUCTION	TESLA	T	Wb/m^2	Derived
MASS	KILOGRAM	kg	kg	Basic
POTENTIAL	VOLT	V	J/C	Derived
POWER	WATT	W	J/s	Derived
PRESSURE	PASCAL	Pa	N/m^2	Derived
RESISTANCE	OHM	Ω	V/A	Derived
TEMPERATURE	KELVIN	K	K	Basic
TIME	SECOND	s	s	Basic

APPENDIX B

Physical Constants

QUANTITY	SYMBOL	VALUE
Angstrom Unit	A°	1 A°=10^{-4} μm=10^{-8} cm
Avogadro's Constant	N_{AVO}	6.02204X10^{23} MOL^{-1}
Boltzmann's Constant	k	1.38066x10^{-23} J/K
Charge of Electron	q_e	1.60218x10^{-19} C
Electron Charge/Mass Ratio	q_e/m_e	1.75880x10^{11} C.kg^{-1}
Electron Rest Mass	m_e	9.1095x10^{-31} kg
Electron Volt	eV	1eV=1.60218x10^{-19} J
Intrinsic Impedance (Vacuum)	η_o	120π (377) Ω
Neutron Rest Mass	m_n	1.67495x10^{-27} kg
Permeability (Vacuum)	μ_o	4πx10^{-7} H/m
Permittivity (Vacuum)	ε_o	8.85418x10^{-12} F/m
Planck's Constant	h	6.62617x10^{-34} J-s
Proton Rest Mass	M_p	1.67264x10^{-27} kg
Speed of Light (Vacuum)	c	2.99792x10^{8} m/s
Thermal Voltage (at 300 K)	V_T=kT/q	0.0259 ≈26 mV

APPENDIX C

The System of Units

In mathematics, since we are dealing with abstract concepts represented by symbols, there is no need for units. However if we are going to assign each of the mathematical symbols a specific physical quantity which is actual and measurable, then we need to know their units. This is an essential part of any scientific methodology where precise units are needed to obtain meaningful and finite answers to the real life problems that are encountered in the process of mankind's survival on this planet.

So all of the scientific discoveries and the corresponding laws are most often cast into mathematical format to facilitate the communication of concepts and are there only to serve and expedite the understanding of the subject for the present and future generations and for no other purpose. Therefore the abstract mathematical formulas are understood clearly when meaningful and practical units of measurements are used.

One of the most useful units of measurement is the SI (international system of units). SI is built upon four fundamental components of the physical universe i.e. Matter, Energy, Space and Time. The fundamental units of measurement are defined as:

a) For mass, we use kilograms (kg).
b) For energy, we use amperes (A)-(unit for current).
c) For space, we use meters (m).
d) For time, we use seconds (s).

These four units in SI are actually borrowed from a former system of measurement, which was commonly referred to as "MKSA system". All other units used in electrical and electronics engineering are derived from these four units and are expressed in terms of meters (m), kilograms (kg), seconds (s) and amperes (A). For example unit for charge is Coulomb (C) which can be expressed as (A-s).

Base and Derived Units

It should be noted that actually there are seven "base units" in the International system of units (SI): 1) meter, 2) kilogram (unit of macro-amount of substance), 3) second, 4) Ampere, 5) Kelvin (unit of temperature), 6) mole (unit of micro-amount of substance) and 7) candela (unit of luminous intensity).

In addition to the seven base units, a second class of SI units exist which are called "derived units". The derived units are formed by combining base units according to the algebraic relations linking the corresponding quantities, and each is usually named after a prominent scientist.

Examples of derived units include Ohm, Farad, Volt, Joule, etc. There exists a third class of SI units, which are called "supplementary units", such as radian, steradian, etc. These three classes of SI units form a coherent and useful system of measurement for all known quantities encountered in the physical universe.

Unit Definitions

We present brief definitions of some of the most commonly used units in physics and electrical engineering in alphabetical order as follows:

1) Ampere (A) – The unit of Electric current defined as the flow of one Coulomb of charge per second. Alternately, it can also be defined as the constant current that would produce a force of 2×10^{-7} Newton per meter of length in two straight parallel conductors of infinite length, of negligible cross section, placed one meter apart in a vacuum.

2) Celsius (°C) - 1/100th of the temperature difference between the freezing point of water (0°C) and boiling point of water (100°C) on the Celsius temperature scale.

3) Coulomb (C) – The unit of electric charge defined as the charge transported across a surface in one second by an electric current of one ampere. An electron has a charge of 1.602×10^{-19} Coulomb.

4) Fahrenheit (°F) – 1/180th of the temperature difference between the freezing point of water (32°F) and boiling point of water (212°F) on the Fahrenheit temperature scale.

5) Farad (F) – The unit of capacitance in the MKSA system of units equal to the capacitance of a capacitor which has a charge of one Coulomb when a potential difference of one volt is applied.

6) Gauss (G) – The unit of Magnetic induction (also called magnetic flux density) in the cgs system of units equal to one line per square centimeter which is the magnetic flux density of one Maxwell per square centimeter or 10^{-4} Tesla.

7) Gilbert (Gi)- The unit of magnetomotive force (mmf) in the cgs system of units, equal to the magnetomotive force of a closed loop of one turn in which there is a current of $1/4\pi$ abampere (1 abampere=10 A). Thus one Gilbert=$10/4\pi$ Ampere-turn.

8) Henry (H) – The unit of self and mutual inductance in the MKSA system of units equal to the inductance of a closed loop that gives rise to a magnetic flux of one Weber for each ampere of current that flows through.

9) Hertz (Hz) – The unit of frequency equal to the number of cycles of a periodic function that occur in one second.

10) Joule (J) – The unit of energy or work in the MKSA system of units, equal to the work performed as the point of application of a force of one Newton moves through one meter distance in the direction of the force.

11) Kelvin (K) – The unit of absolute temperature, equal to a value of 1/273.16 of the absolute temperature of the triple point of water (which is a particular temperature, 273.16 K, and pressure point at which three different phases of water, i.e., vapor, liquid and ice can coexist in equilibrium).

12) Maxwell (Mx) – The unit for Magnetic Flux in the cgs system of units, equal to 10^{-8} Weber.

13) Newton (N) – The unit of force in MKSA system of units equal to the force that imparts an acceleration of one meter per second squared to a mass of one kilogram.

14) Oersted (Oe) – The unit of Magnetic Field in the cgs system of units equal to the field strength at the center of a plane circular coil of one turn and 1-cm radius when there is a current of $10/2\pi$ ampere in the coil.

15) Ohm (Ω) – The unit of resistance in the MKSA system of units equal to the resistance between two points on a conductor through which a current of one ampere flows as a result of a potential difference of one volt applied between the two points.

16) Siemens (S) – (also called mho, inverse of Ohm) – The unit for conductance, susceptance and admittance in the MKSA system of units and is equal to the reciprocal of the resistance of an element that has a resistance of one ohm.

17) Tesla (T) – The unit of magnetic field in the MKSA system of units equal to one Weber per square meter.

18) Volt (V) – The unit of potential difference (or electromotive force) in the MKSA system of units equal to the potential difference between two points for which 1 Coulomb of charge will do one joule of work in going from one point to the other.

19) Watt (W) – The unit of power in MKSA system of units defined as the work of one joule done in one second.

20) Weber (Wb) – The unit of Magnetic Flux in the MKSA system of units equal to the magnetic flux which, linking a circuit of one turn, produces an electromotive force of one volt when the flux is reduced to zero at a uniform rate in one second.

Note: "cgs" is Centimeter-Gram-Second system of units.

Glossary of Technical Terms

The following glossary supplements the presented materials in the text, but does not replace the use of an unabridged technical dictionary, which is a must for mastery of sciences.

Absolute
a) That which is without reference to anything else and thus not comparative or dependent upon external conditions for its existence (opposed to relative), b) That which is free from any limitations or restrictions and is thus unconditionally true at all times.

Absolute Temperature Scale
A scale with which temperatures are measured relative to absolute zero (the temperature of –273.15 °C or –459.67 °F or 0 K). The absolute temperature scale leads to the absolute temperatures, which are: a) The temperature in Celsius degrees, relative to –273.15 °C (giving rise to the Kelvin scale), and b) The temperature in Fahrenheit degrees, relative to –459.67 °F (giving rise to the Rankine scale). At absolute zero temperature, molecular motion theoretically vanishes and a body would have no heat energy. The absolute zero temperature is approachable but never attainable. *See also* **temperature**.

Ampere (A)
The unit of electric current defined as the flow of one Coulomb of charge per second. Alternately, it can also be defined as the constant

current that would produce a force of $2x10^{-7}$ Newton per meter of length in two straight parallel conductors of infinite length, and of negligible cross section, placed one meter apart in a vacuum.

Analog
Pertaining to the general class of devices or circuits in which the output varies as a continuous function of the input.

Application Mass
All of the related masses that are connected and/or obtained as a result of the application of a science. This includes all physical devices, machines, experimental setups, and other physical materials that are directly or indirectly derived from and are a result of the application. In this book when we say application mass, we really mean "technical application mass." See also Generalized application mass, Technical application mass and personalized application mass.

Average Power
The power averaged over one cycle.

Axiom
A self-evident truth accepted without proof.

B

Bidirectional
Responsive in both directions.

Bilateral
Having a voltage-current characteristic curve that is symmetrical with respect to the origin. If a positive voltage produces a positive current magnitude, then an equal negative voltage produces a negative current of the same magnitude.

C

Capacitor
A device consisting essentially of two conducting surfaces separated by an insulating material (or a dielectric) such as air, paper, mica, etc., that can store electric charge.

Celsius (°C)
1/100th of the temperature difference between the freezing point of water (0°C) and the boiling point of water (100°C) on the Celsius temperature scale given by:

$$T(°C)=T(K)-273.15=\frac{5}{9}\{T(°F)-32\}.$$

Charge
A basic property of elementary particles of matter (electrons, protons, etc.) that is capable of creating a force field in its vicinity. The built-in force field is a result of stored electric energy.

Chip
A single substrate upon which all the active and passive circuit elements are fabricated using one or all of the semiconductor techniques of diffusion, passivation, masking, photoresist, epitaxial growth, etc.

Circuit
The interconnection of a number of devices in one or more closed paths to perform a desired electrical or electronic function.

Classical Mechanics (Also Called Classical Physics, Non-Quantized Physics or Continuum Physics)
Is the branch of physics based on concepts established before quantum physics, and includes materials in conformity with Newton's mechanics and Maxwell's electromagnetic theory.

Code
Any unique set of data or combination of symbols, signals, actions, etc., that will open a passageway to a hidden location (such as a vault), or unlock the pattern of behavior of some puzzling and mysterious data, thus revealing the information behind the mystery.

Communication Principle (Also Called The Universal Communication Principle)
A fundamental concept in life and livingness that is intertwined throughout the entire field of sciences that states for communication to take place between two or more entities, three elements must be present: a source point, a receipt point, and an imposed space or distance between the two.

Component
A packaged functional unit consisting of one or more circuits made up of devices, which in turn may be part of an operating system or subsystem.

Conductor
a) A material that conducts electricity with ease, such as metals, electrolytes, and ionized gases; b) An individual metal wire in a cable, insulated or un-insulated.

Current
Net transfer of electrical charges across a surface per unit time, usually represented by (I) and measured in Ampere (A). Current density (J) is current per unit area.

D

DC (Also Called Direct Current)
A current which always flows in one direction (e.g., a current delivered by a battery).

Device
A single discrete conventional electronic part such as a resistor, a transistor, etc.

Dichotomy
Two things or concepts that are sharply or distinguishably opposite to each other.

Die (Also Called Chip)
A single substrate on which all the active and passive elements of an electronic circuit have been fabricated. This is one portion taken from a wafer bearing many chips, but it is not ready for use until it is packaged and provided with terminals for connection to the outside world.

Dielectric
A material that is a non-conductor of electricity. It is characterized by a parameter called *dielectric constant* or *relative permittivity* (ε_r).

Digital
Circuitry in which data-carrying signals are restricted to either of two voltage levels.

Discovery
The gaining of knowledge about something previously unknown.

Discrete Device
An individual electrical component such as a resistor, capacitor, or transistor as opposed to an integrated circuit that consists of several discrete components.

Dual
Two concepts, energy forms or physical things that are of comparable magnitudes but of opposite nature, thus becoming counterpart of each other.

Duality Theorem
States that when a theorem is true, it will remain true if each quantity and operation is replaced by its dual quantity and operation. In circuit theory, the dual quantities are “voltage and current” and “impedance and admittance.” The dual operations are “series and parallel” and “meshes and nodes.”

Electric Charge (or Charge)
(Microscopic) A basic property of elementary particles of matter (e.g., electron, protons, etc.) that is capable of creating a force field in its vicinity. This built-in force field is a result of stored electric energy. (Macroscopic) The charge of an object is the algebraic sum of the charges of its constituents (such as electrons, protons, etc.), and may be zero, a positive or a negative number.

Electric Current (or Current)
The net transfer of electric charges (Q) across a surface per unit time.

Electric Field
The region about a charged body capable of exerting force. The intensity of the electric field at any point is defined to be the force that would be exerted on a unit positive charge at that point.

Electric Field Intensity
The electric force on a stationary positive unit charge at a point in an electric field (also called *electric field strength*, *electric field vector*, and *electric vector*).

Electricity
Is a form of energy, which can be subdivided into two major categories: a) Electrostatics, and b) Electrokinetics.

Electromagnetic (EM) Wave
A radiant energy flow produced by oscillation of an electric charge as the source of radiation. In free space and away from the source, EM rays of waves consist of vibrating electric and magnetic fields that move at the speed of light (in vacuum), and are at right angles to each other and to the direction of motion. EM waves propagate with no actual transport of matter, and grow weaker in amplitude as they travel farther in space. EM waves include radio, microwaves, infrared, visible/ultraviolet light waves, X-ray, gamma rays, and cosmic rays.

Electron
A stable elementary particle of matter, which carries a negative electric charge of one electronic unit equal to q= -1.602×10^{-19} C and has a mass of about 9.11×10^{-31} kg and a spin of ½.

Electronics
The study, control, and application of the conduction of electricity through different media (e.g., semiconductors, conductors, gases, vacuum, etc.).

Elementary Particle
A particle, which can not be described as a compound of other particles and is thus one of the fundamental constituents of all matter (e.g. electron, proton, etc.).

Energy
The capacity or ability of a body to perform work. Energy of a body is either potential motion (called *potential energy*) or due to its actual motion (called *kinetic energy*).

Entropy
A measure of the degree of disorder, randomness, and uncertainty in a system caused by the random change of particle's location in a created space, random alteration of data and facts pertaining to a certain event, etc. The entropy of a closed system always increases, because of the inherent postulate of "change" in the physical universe (see the original postulates).

Fahrenheit (°F)
1/180th of the temperature difference between the freezing point of water (32°F) and the boiling point of water (212°F) on the Fahrenheit temperature scale.

$$T(°F) = T(°R) - 459.67 = \frac{9}{5}T(°C) + 32$$

Where °R and °C are symbols for degrees Rankin and Celsius, respectively.

Field
An entity that acts as an intermediary agent in interactions between particles, is distributed over a region of space, and whose properties are a function of space and time, in general.

Field Theory
The concept that, within a space in the vicinity of a particle, there exists a field containing energy and momentum, and that this field interacts with neighboring particles and their fields.

Flow
The passage of particles (e.g., electrons, etc.) between two points. Example: electrons moving from one terminal of a battery to the other terminal through a conductor. The direction of flows are from higher to lower potential energy levels.

Force
That form of energy that puts an unmoving object into motion, or alters the motion of a moving object (i.e., its speed, direction or both). Furthermore, it is the agency that accomplishes work.

Frequency
The number of complete cycles in one second of a repeating quantity, such as an alternating current, voltage, electromagnetic waves, etc.

Gain
The ratio that identifies the increase in signal or amplification that occurs when the signal passes through a circuit.

Ground
(a) A metallic connection with the earth to establish zero potential (used for protection against short circuit); (b) The voltage reference point in a circuit. There may or may not be an actual connection to earth but it is understood that a point in the circuit said to be at ground potential could be connected to earth without disturbing the operation of the circuit in any way.

Generalized Application Mass (G.A.M.)
In general, is any created space, which contains created energies and created matter of any form, shape or size existing as a function of time. In simple terms, generalized application mass is any matter and energy, condensed and packaged into an object form, which exists in a time-stream (from its inception to now). The generalized concept of application mass includes the entire mechanical space containing all energies and matter such as electrons, atoms, molecules and all the existing gigantic masses of planets, stars, galaxies, which are not the direct byproduct of Man's sciences.

Hertz (Hz)
The unit of frequency equal to the number of cycles of a periodic function that occur in one second.

Hole
A vacant electron energy state near the top of the valence band in a semiconductor material. It behaves as a positively charged particle having a certain mass and mobility. It is the dual of electron, unlike a proton which is the dichotomy of an electron.

Hypothesis
An unproven theory or proposition tentatively accepted to explain certain facts or to provide a basis for further investigation.

I

Inductance (L)
The inertial property of an element (caused by an induced reverse voltage), which opposes the flow of current when a voltage is applied; it opposes a change in current that has been established.

Inductor
A conductor used to introduce inductance into an electric circuit, normally configured as a coil to maximize the inductance value.

Input
The current, voltage, power, or other driving force applied to a circuit or device.

Insulator
A material in which the outer electrons are tightly bound to the atom and are not free to move. Thus, there is negligible current through the material when a voltage is applied.

Integrated Circuit (IC)
An electrical network composed of two or more circuit elements on a single semiconductor substrate.

Isolation
Electrical separation between two points.

J

Joule (J)
The unit of energy or work in the MKSA system of units, which is equal to the work performed as the point of application of a force of one Newton moves the object through a distance of one meter in the direction of the force.

K

Kelvin (K)
The unit of measurement of temperature in the absolute scale (based on Celsius temperature scale), in which the absolute zero is at –273.15 °C. It is precisely equal to a value of 1/273.15 of the absolute temperature of the triple point of water, being a particular pressure and temperature point, 273.15 K, at which three different phases of water (i.e., vapor, liquid, and ice) can coexist at equilibrium. *See also* **temperature**.

Kinetic
(*Adjective*) Pertaining to motion or change. (*Noun*) Something which is moving or changing constantly such as a piece of matter.

Kinetic Energy (K.E.)
The energy of a particle in motion. The motion of the particle is caused by a force on the particle.

Knowledge
Is a body of facts, principles, data, and conclusions (aligned or unaligned) on a subject, accumulated through years of research and investigation, that provides answers and solutions in that subject.

L

Law
An exact formulation of the operating principle in nature observed to occur with unvarying uniformity under the same conditions.

Law of Conservation of Energy (Excluding All Metaphysical Sources of Energy)
This fundamental law simply states that any form of energy in the physical universe can neither be created nor destroyed, but only converted into another form of energy (also known as the principle of conservation of energy).

Light Waves
Electromagnetic waves in the visible frequency range, which ranges from 400 nm to 770 nm in wavelength.

M

Magnet
A piece of ferromagnetic or ferromagnetic material whose internal domains are sufficiently aligned so that it produces a considerable net magnetic field outside of itself and can experience a net torque when placed in an external magnetic field.

Magnetic Field
The space surrounding a magnetic pole, a current-carrying conductor, or a magnetized body that is permeated by magnetic energy and is capable of exerting a magnetic force. This space can be characterized by magnetic lines of force.

Man
Homo sapiens (literally, the knowing or intelligent man); mankind.

Mathematics
Mathematics are short-hand methods of stating, analyzing, or resolving real or abstract problems and expressing their solutions by symbolizing data, decisions, conclusions, and assumptions.

Matter
Matter particles are a condensation of energy particles into a very small volume.

Mechanics
The totality of the three categories of application mass: a) Generalized application Mass; b) Technical application mass, and c) Personalized application mass. See also classical mechanics and quantum mechanics.

Microelectronics
The body of electronics that is associated with or applied to the realization of electronic systems from extremely small electronic

Microwaves
Waves in the frequency range of 1 GHz to 300 GHz.

Millimeter Wave
Electromagnetic radiation in the frequency range of 30 to 300 GHz, corresponding to wavelength ranging from 10 mm to 1 mm.

Model
A physical (e.g., a small working replica), abstract (e.g., a procedure) or a mathematical representation (e.g., a formula) of a process, a device, a circuit, or a system and is employed to facilitate their analysis.

N

Natural Laws
A body of workable principles considered as derived solely from reason and study of nature.

Neutron
One of uncharged stable elementary particles of an atom having the same mass as a proton. A free neutron decomposes into a proton, an electron, and a neutrino. A neutrino is a neutral uncharged particle but is an unstable particle since it has a mass that approaches zero very rapidly (a half-life of about 13 minutes).

Newton (N)
The unit of force in MKSA system of units equal to the force that imparts an acceleration of one m/s^2 to a mass of one kilogram.

Newton Laws of Motion
Three fundamental principles (called Newton's first, second, and third laws of motion), which form the basis of classical mechanics (also called Newtonian mechanics) and have proved valid for all mechanical problems, not involving speeds comparable with the speed of light and not involving atomic or subatomic particles.

Newton's First Law of Motion
The law that a particle not subjected to external forces remains at rest or moves with constant speed in a straight line.

Newton's Second Law of Motion
The law that the acceleration of a particle is directly proportional to the resultant net external force acting on the particle and is inversely proportional to the mass of the particle.

Newton's Third Law of Motion
The law that, if two particles interact, the force exerted by the first particle on the second particle (called the action force) is equal in magnitude and opposite in direction to the force exerted by the second particle on the first particle (called the reactive force). This is also called the law of action and reaction.

Noise

Random unwanted electrical signals that cause unwanted and false output signals in a circuit.

Nomenclature
The set of names used in a specific activity or branch of learning; terminology.

Nonlinear
Having an output that does not rise and fall in direct proportion to the input.

Nucleus
The core of an atom composed of protons and neutrons, having a positive charge equal to the charge of the number of protons that it contains. The nucleus contains most of the mass of the atom, pretty much like the sun containing most of the mass of the solar system.

Occam's (or Ockham's) Razor Doctrine
A principle that assumptions introduced to explain a thing must not be multiplied beyond necessity. In simple terms, it is a principle stating that the simplest explanation of a phenomenon, which relates all of the facts, is the most valid one. Thus by using the Occam's razor doctrine a complicated problem can be solved through the use of simple explanations, much like a razor cutting away all undue complexities (after William of Occam, an English philosopher, 1300-1349, who made a great effort to simplify scholasticism).

Original Postulates
A series of exact postulate (space, energy, change) that have gone into the construction of the physical universe. See the primary postulates.

Oscillator
An electronic device that generates alternating-current power at a frequency determined by constants in its circuits.

Output
The current, voltage, power, or driving force delivered by a circuit or device.

P

Particle
Any tiny piece of matter, so small as to be considered theoretically without magnitude (i.e., zero size), though having mass, inertia and the force of attraction. Knowing zero size is an absolute and thus impossible in the physical universe, practical particles range in diameter from a fraction of angstrom (as with electrons, atoms and molecules) to a few millimeters (as with large rain drops).

Passive
A component that may control but does not create or amplify electrical energy.

Perfect Conductor
Is a conductor having infinite conductivity or zero resistivity.

Personalized Application Mass (P.A.M.)
Is the category of application mass, which has been created and is based solely upon the viewpoint's own postulates and considerations. Examples of this category include such things as one's own customized possessions, any piece of artwork or music, one's own body characteristics (such as hairdo, clothing, shape, etc.), a book's layout or cover design, so on and so forth. see also application mass, Technical application mass, and Generalized application mass.

Phase
The angular relationship of a wave to some time reference or other wave.

Physical Universe (Also Called The Material Universe or The Universe)
Is a universe based upon three postulates, called original postulates (space, energy and change) and has four main components (matter, energy, space and time).

Postulate
a) (NOUN) is an assumption or assertion set forth and assumed to be true unconditionally and for all times without requiring proof; especially as a basis for reasoning or future scientific development;

b) (VERB) To put forth or assume a datum as true or exist without proof.

Potential Energy (P.E.)
Any form of stored energy that has the capability of performing work when released. This energy is due to the position of particles relative to each other.

Primary Postulates
A series of four postulates derived from the original postulates. These postulates are responsible for the four basic components of the physical universe: matter, energy, created space, and mechanical time. See the original postulates.

Principle
A rule or law illustrating a natural phenomenon, operation of a machine, the working of a system, etc.

Propagation
The travel of a wave through a medium.

Proton
An elementary particle, which is one of the three basic subatomic particles, with a positive charge equivalent to the charge of an electron ($q = +1.602 \times 10^{-19}$ C) and has a mass of about 1.67×10^{-27} kg with a spin of ½. Proton together with neutron is the building block of all atomic nuclei.

Pulse
A variation of a quantity, which is characterized by a rise to a certain level (amplitude), a finite duration, and a decay back to the normal level.

Pyramid of Knowledge
Workable knowledge forms a pyramid, where from a handful of common denominators efficiently expressed by a series of basic postulates, axioms and natural laws, which form the foundation of a science, an almost innumerable number of devices, circuits and systems can be thought up and developed. The plethora of the mass of devices, circuits and systems generated is known as the "application mass", which practically approaches infinity in sheer number.

Q

Quantum Mechanics (Also Called Quantum Physics or Quantum Theory)
Is the study of atomic structure which states that an atom or molecule does not radiate or absorb energy continuously. Rather, it does so in a *series of steps, each step being the emission or absorption of an amount of energy packet (E) called a quantum.* Quantum physics is the modern theory of matter, electromagnetic radiation and their interaction with each other. It differs from classical physics in that it generalizes and supersedes it, mainly in the realm of atomic and subatomic phenomena.

Quark
A hypothetical basic particle having a fraction of charge of an electron (such as 1/3 or 2/3) from which many of the elementary particles (such as electrons, protons, neutrons, mesons, etc.) may be built up theoretically. No experimental evidence for the actual existence of free quarks has been found.

R

Radio Frequency (RF)
Any wave in the frequency range of a few kHz to 300 MHz, at which coherent electromagnetic radiation of energy is possible.

Rankine (°R)
The unit of measurement of temperature in the absolute scale (based on Fahrenheit temperature scale), in which the absolute zero is at -459.67 °F. *See also* **temperature**.

Resistor
A lumped bilateral and linear element that impedes the flow of current, through it when a potential difference, is imposed between its two terminals.

S

Science
A branch of study concerned with establishing, systematizing, and aligning laws, facts, principles, and methods that are derived from hypothesis, observation, study and experiments.

Semiconductor
A material having a resistance between that of conductors and insulators, and usually having a negative temperature coefficient of resistance.

Signal
A sign given so as to convey a command, direction or warning; more specifically in electricity, an electrical quantity (such as a current or voltage) that can be used to convey information for communication, control, etc.

Space (Also Called Created Space)
The continuous three-dimensional expanse extending in all directions, within which all things under consideration exist.

Standing Wave
A standing, apparent motionless-ness, of particles causing an apparent no out-flow, no in-flow. A standing wave is caused by two energy flows, impinging against one another, with comparable magnitudes to cause a suspension of energy particles in space, enduring with a duration longer than the duration of the flows themselves.

Static
(*Adjective*) Pertaining to no-motion or no-change. (*Noun*) Something which is without motion or change such as truth (an abstract concept). In physics, one may consider a very distant star (a physical universe object) a static on a short term basis, but it is not totally correct because the distant star is moving over a long period of time, thus is not truly a static but only an approximation, or a physical analogue of a true static.

Subjective Time
Is the consideration of time in one's mind, which can be a nonlinear or linear quantity depending on one's viewpoint.

Symbiont
An organism living in a state of association and interdependence with another kind of organism, especially where such association is of mutual advantage, such as a pet. Such a state of mutual interdependence is called "symbiosis."

T

Technical Application Mass (T.A.M.)
Is the category of man-made application mass that is produced directly as a result of application of a science using its scientific postulates, axioms, laws and other technical data. Examples include such things as a television set, a computer, an automobile, a power generator, a telephone system, a rocket, etc. See also Application mass, Personalized application mass, and Generalized application mass.

Technology
The application of a science for practical ends.

Temperature
The degree of hotness or coldness measured with respect to an arbitrary zero or an absolute zero, and expressed on a degree scale. Examples of arbitrary-zero degree scales are Celsius scale (°C) and Fahrenheit scale (°F); and examples of absolute-zero degree scales are Kelvin degree scale (based on Celsius degree scale) and Rankine degree scale (based on Fahrenheit degree scale).

Theorem
A proposition that is not self-evident but can be proven from accepted premises and therefore, is established as a principle.

Theory
An explanation based on observation and reasoning, which explains the operation and mechanics of a certain phenomenon. It is a generalization reached by inference from observed particulars and

implies a larger body of tested evidence and thus a greater degree of probability. It uses a hypothesis as a basis or guide for its observation and further development.

Thermodynamics

The branch of physics that deals with the relations between heat and other forms of energy (especially mechanical energy), and with the governing laws regarding the conversion of one into the other. Applications of thermodynamics include steam engines, refrigeration systems, cooling and heating engines, etc.

Time (Also Called Mechanical Time or Objective Time)
That characteristic of the physical universe at a given location that orders the sequence of events on a microscopic or macroscopic level. It proceeds from the interaction of matter and energy and is merely an "index of change," used to keep track of a particle's location. The fundamental unit of time measurement is supplied by the earth's rotation on its axis while orbiting around the sun. It can also alternately be defined as the co-motion and co-action of moving particles relative to one another in space. See also subjective time.

Torque
A force that tends to produce rotation or twisting.

U

Unidirectional
Flowing in only one direction (e.g., direct current).

Unilateral
Flowing or acting in one direction only causing a non-reciprocal characteristic.

Universal Communication Principle (Also Called Communication Principle)
A fundamental concept in life and livingness that is intertwined throughout the entire field of sciences that states for communication to take place between two or more entities, three elements must be present: a source point, a receipt point, and an imposed space or distance between the two.

Universe (Derived From Latin Meaning "Turned Into One", "A Whole)
Is the totality or the set of all things that exist in an area under consideration, at any one time. In simple terms, it is an area consisting of things (such as ideas, masses, symbols, etc.) that can be classified under one heading and be regarded as one whole thing.

Viewpoint
Is a point on a mental plane from which one creates (called postulating viewpoint) or observes (called observing viewpoint) an idea, an intended subject or a physical object.

Volt (V)
The unit of potential difference (or electromotive force) in the MKSA system of units equal to the potential difference between two points for which one Coulomb of charge will do one joule of work in going from one point to the other.

Voltage
Voltage or potential difference between two points is defined to be the amount of work done against an electric field in order to move a unit charge from one point to the other.

Voltage Source
The device or generator connected to the input of a network or circuit.

Watt (W)
The unit of power in MKSA system of units defined as the work of one joule done in one second.

Wave
A disturbance that propagates from one point in a medium to other points without giving the medium as a whole any permanent displacement.

Wave Propagation
The travel of waves (e.g., electromagnetic waves) through a medium.

Wavelength
The physical distance between two points having the same phase in two consecutive cycles of a periodic wave along a line in the direction of propagation.

Work
The advancement of the point of application of a force on a particle.

Recommended Resources

1. Boas, M. *The Scientific Renaissance*, New York: Harper, 1965.
2. Buckley, H. *A Short History of Physics*, London: Methuen, 1927.
3. Cajori, F., *A History of Physics*, New York: Dover, 1962.
4. Dampier, W. C. *A History of Science,* Cambridge: Cambridge University Press, 1966.
5. Eker, T. Harv, *Secrets of the Millionaire Mind*, New York: HarperCollins Publishers, Inc., 2005.
6. Hart, I. *Makers of Science*, New York: Oxford University Press, 1923.
7. Hawking, S. W. *A Brief History of Time*, New York: Bantam Books, 1988.
8. Hubbard, L. Ron. *Understanding the E-meter*, Bridge Publications, 1982.
9. Mackenzie, A. E. E. *The Major Advancements in Science,* Cambridge: Cambridge University Press, 1960.
10. Marion, J. V. *A Universe of Physics*, New York: Wiley, 1970.
11. Pledge, H. T. *Science Since 1500.* London: Science Museum, 1966.
12. Radmanesh, M. M. *The Gateway to Understanding: Electrons to Waves and Beyond*, Bloomington: AuthorHouse, 2005
13. Radmanesh, M. M. *The Gateway to Understanding: Electrons to Waves and Beyond WORKBOOK*, Bloomington: AuthorHouse, 2005.

14. Radmanesh, M. M. *Radio Frequency and Microwave Electronics Illustrated*, Upper Saddle River: Prentice Hall, 2001.
15. Radmanesh, M. M. *Obstacles to Comprehension of Engineering Sciences*, Annual IEE Engineering Conference Digest, Sacramento, CA, April 1992.
16. Radmanesh, M. M. *Creativity in Engineering Education for Higher Student Retention*, ASEE Pacific Southwest (PSW) Engineering Conference Digest, Flagstaff, AZ, October 1993.
17. Runes, D. D. *A Treasury of World Science*, New York: Philosophical Library, 1961.
18. Schwartz, G., and P. W. Bishop. *Moments of Discovery*, New York: Basic Books, 1962.
19. Taton, R. *History of Science*, Vols. I-IV. New York: Basic Books, 1964.
20. Wightman, W.P.D. *The Growth of Scientific Ideas*, Yale University press, New Haven, Conn., 1969.

Index

About the Author

Matthew M. Radmanesh received his BSEE degree from Pahlavi University in electrical engineering in 1978, his MSEE and Ph.D. degrees from the University of Michigan, Ann Arbor, in Microwave Electronics and Electro-Optics in 1980 and 1984, respectively.

He has worked in academia for Kettering University (formerly GMI Engineering & Manage-ment) and in industry for Hughes Aircraft Co., Maury Microwave Corp. and Boeing Aircraft Co. He is currently a faculty member in the electrical and computer engineering department at California State University, Northridge, CA.

Dr. Radmanesh is a senior member of IEEE, Eta Kappa Nu Honor society and a past president (three years) of the SFV Chapter of the IEEE Microwave Theory and Technique (MTT) society. His many years of experience in both microwave industry and academia have led to over 40 technical papers in national and international journals and several design handbooks in microwave engineering and in solid state devices and integrated circuit engineering.

His current research interests include design of RF and Microwave devices and circuits, millimeter-wave circuit applications, photonic engineering as well as engineering education. He received the distinguished lecturer award at the 1994 IEEE international Microwave Symposium and was awarded twice by IEEE LA council for his contributions to the MTT society (1994, 1995). He also received two awards for commitment and dedication to education from IEEE in 2002 and 2003.

Dr. Radmanesh won the MPD divisional award while at Hughes Aircraft Co. for his pioneering work in the development and design of solid state millimeter wave noise sources in Ka- and V-band, and a similar award for his outstanding contributions to the HERF project from Boeing Aircraft Co. He holds two patents for his pioneering work and novel designs of two millimeter-wave noise sources.

Dr. Radmanesh authored a popular book entitled "*The Gateway to Understanding: Electrons to Waves and Beyond,*" accompanied by a comprehensive WORKBOOK, both published by AuthorHouse in 2005 as well as another book entitled *"Radio Frequency and Microwave Electronics Illustrated"* published by Prentice Hall in 2001. His hobbies include chess, philosophy, soccer and tennis.

Dr. Radmanesh intends to bring about a higher level of understanding in the scientific community and the society at large, about the basic principles of life and livingness of which the knowledge about sciences and engineering is but a subset.

Quick Order Form

Fax orders: Send this filled out form to: 818-700-8933.
Telephone orders: Call 1(888) 280-7715 toll free. Have your credit card ready.
Email orders: admin@krcbooks.com, or mattradman@sbcglobal.net
Postal Orders: KRC, Matthew Radmanesh,
PO Box 280188, Northridge, CA 91328-0188,
USA Telephone: 888-280-7715

Please send the following books, disks or reports. I understand that I may return any of them in the *original condition* for a full refund—for any reason, no questions asked.

____
____
____
____

Please send more FREE information on:
o Other Books o Consulting o Speaking/Seminars

Name:____

Address:____

City: ____________________State: _____Zip: ______

Telephone:______________________________________

Email address:__________________________________

Sales tax: Please add 8.25% for products shipped to California addresses.

Shipping by air:
U.S.: $4.00 for first book or disk and $2.00 for each additional product.
International: $9.00 for first book or disk; $5.00 for each additional product (estimate).

Payment: o Check o Money order
Make the check or money order payable to: KRC

Other Books By The Author :

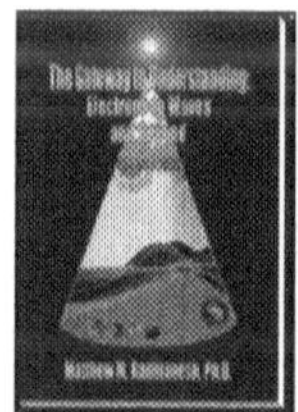

The Gateway to Understanding: Electrons to Waves and Beyond, AuthorHouse, 2005.

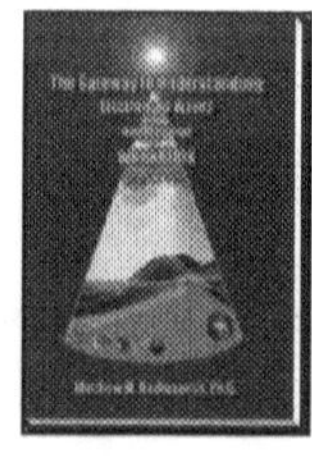

The Gateway to Understanding: Electrons to Waves and Beyond WORKBOOK, AuthorHouse, 2005.

Radio frequency and Microwave Electronics Illustrated, Prentice Hall, 2001.

For more information or to order any of the books, please visit:

www.KRCbooks.com

www.ingramcontent.com/pod-product-compliance
Lightning Source LLC
Chambersburg PA
CBHW031815170526
45157CB00001B/61

* 9 7 8 1 4 2 5 9 1 5 9 9 5 *